Fundamentals of Calculus and Probability

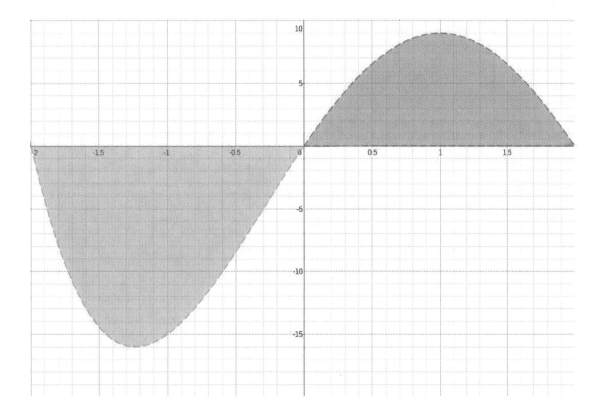

Timothy C. Kearns

AuthorHouse™ LLC
1663 Liberty Drive
Bloomington, IN 47403
www.authorhouse.com
Phone: 1-800-839-8640

Published by AuthorHouse   06/26/2014

ISBN: 978-1-4969-0482-9 (sc)
ISBN: 978-1-4969-0481-2 (e)

Dedicated to my family,
my parents Albert W. and Amy S. Kearns,
my brother John S. Kearns,
and my nephews John Q. and Matthew.

About the Author

My name is Timothy Clayton Kearns. I was born in Arlington, Va. USA, and except for my college years, I have lived in northern Virginia all my life. I have spent the last 35 years studying formally and informally the subjects of statistics, mathematics, elementary physics, the history of science, and many other subjects, to various degrees. In June 1983, in Blacksburg, Va, I graduated (cum laude) from Virginia Polytechnic Institute and State University, with a Bachelor of Science degree in statistics and a strong minor in mathematics. I have also completed some graduate level study in mathematics and statistics at George Mason University in Fairfax, Va, from 1985 to 1989. I have taught calculus as a graduate assistant at GMU and have spent the last 11 years as a self-employed math tutor in the northern Va. area, tutoring students from the middle school level to the PhD level. This book is a serious attempt to satisfy some desire within me to help others appreciate and understand some important fundamental parts of mathematical and statistical analysis, subjects which have given me much pleasure over the years.

Timothy C. Kearns
Springfield, Va.
June 19, 2014

Table of Contents

Introduction

Mathematical analysis has been a lifelong passion of mine and it is my goal in this book to share some of the more fundamental, important and interesting parts of this subject with serious students that intend to pursue advanced study in the mathematical sciences, or related areas that require advanced math. This book provides a solid introduction to set theory, in particular infinite sets and advanced topics concerning the real number system, a good review of elementary calculus, a good introduction to probability theory, and some elementary statistics. It is my belief that many of the important ideas in the study of infinite sets and the real numbers, which are usually taught at the graduate level, can be introduced to students before graduate school. This book will provide a solid understanding of the modern mathematical foundations necessary for advanced study in mathematics and statistics. To get the most out of this book it is assumed that the reader has completed at least one or two college level courses in mathematics and statistics, including calculus. This is not a textbook, but it is a structured informal treatment of mathematics, probability, and elementary statistics, that contains an abundance of examples to illustrate the concepts. It is hoped that those readers, intellectually serious, and planning to do more advanced work in mathematics and its applications, will profit from reading this book.

Timothy C. Kearns
Springfield, Va.
June 19, 2014

CHAPTER 1:
A FEW WORDS ABOUT SCIENCE

(1) **THE SCIENTIFIC WAY**

This is a book about some important parts of modern mathematical and statistical analysis, which has been developed during the past 300-400 years or so. Many people knowledgeable about the history of science would agree that the emergence of mathematical knowledge has transformed our world, of course during antiquity, but most particularly in the last few centuries. Primarily what mathematics and statistics has done is allow us to discover truths about the world through our science, which has removed much uncertainty about the world we live in, therefore allowing us to have a much better understanding of our world and a greater mastery of it. This does not mean of course that the change that it has brought has all been good.

Human civilization has existed for about 10,000 years or so, starting in the middle east, northern africa, and parts of southern asia. Before this time we were most likely loosely organized hunter-gatherers. Uncertainty for ancient humans was largely centered around the perils of the wild: Will we eat today or will we be eaten? Will a neighboring clan attack us? Most of the world around ancient man was a complete mystery. There was no understanding of the weather, or the workings of the celestial bodies that dominated day and night. We didn't understand the composition of the ground beneath our feet or the air that filled our lungs. Life expectancy was probably about 40 years in most parts of the world. We were simply another part of the animal kingdom and the natural world. Uncertainty for modern man still comes from lack of knowledge about the physical world, but there is a new element, there is now a lot of uncertainty that is the result of our own making. Will greenhouse gases and other pollution destroy our food supply? Will another country attack us with nuclear weapons? How can we get this computer chip to work three times faster?

Today we have developed ways to use probability to quantify some of our unknowns, to quantify uncertainty. There is a 60% chance of rain today, a person will on average be involved in four car accidents in their life, there is approximately a 34.5% chance that an electron will be in a certain region of space, and the chance of a 40 year old man surviving a certain disease is 90%, etc. We know a lot more about ourselves and the world around us, and it is advances in modern science and mathematics that has made this possible. Among the most important subjects in modern history that have been developed from mathematics are probability and statistics, and many parts of physical science. This book deals with some important topics in mathematics, probability, and statistics which are fundamental to science. Statistical science is a very extensive subject and we will leave the greater part of the subject of statistics for those more comprehensive textbooks which are out there, written by people that know much more about statistics than I do. I will devote only one chapter, the final chapter, to some of the basic ideas of statistics, which I see as an application of probability theory.

Concerning the history of religious faith and science, faith-based belief systems and religions involving god(s) or other supernatural entities have emerged along with civilization, and have existed from the distant past to the modern day. They have always been an important part of the way that we see the world, but they do not involve mathematics. Any belief system not based on mathematics has only limited explanatory capability. This is the strength and advantage of modern science over religious belief, in as far as explaining the natural world is concerned. This has led to a situation where the emergence of modern science has sometimes been at odds with religion. This is particularly true in more recent times, mostly the last 150-200 years, with new understandings in geology and the theory of biological evolution. My belief is that the natural world, which I believe to be God, and possessing some kind of creative energy, necessarily follows a mathematical design. That is, the natural world is constrained by the laws of mathematics. All things follow the laws of mathematics and are constrained by them because we live in a universe of logical rules and consequences, which are immutable. This is perhaps the essence of God. If there were no immutable logical laws, then the world would be without patterns or rules and there would be nothing. It seems to me that something must exist, because a state of nothingness is itself something. An immutable logic, and hence mathematics, is the underpinning of everything. So we can say that design is the essence of the natural world and it is expressed to man in the language of the mathematics and science that we have discovered.

The first glimmerings of true science started in and around ancient Greece and other parts of the mediterranean world, about 2500 years ago. This is believed to be the time of the first Greek philosophers and scientific thinkers, ushering in new ways of thinking about and describing the world. These times are correlated with the start of significant advancements and developments in mathematics, mainly geometry and the development of rigorous mathematical proof. Basic reasoning with numbers, and calculations, and geometry had existed in some cultures for thousands of years. However, with the ancient greeks there was a significant rise in philosophical and scientific thinking and rigorous mathematical thinking. There are many notable men of those times that are known to historians. Surely most people have heard of the great philosophers Socrates, Plato, and Aristotle, and the mathematicians Pythagoras, Euclid, and Archimedes, among many others.

Greek science and mathematics flourished for hundreds of years in the Mediterranean world, but the dark and middle ages eventually descended over the European and Mediterranean world with the rise and fall of political, military, and religious empires. This brought a virtual halt to science in the ancient world of the Mediterranean and Europe, until Asian and Arabic peoples of the late middle ages brought new ideas in arithmetic and algebra (with the modern positional number system that we use today) into the Mediterranean areas and Europe. Once Asian and Arabic arithmetic methods spread into Europe in the late middle ages, the stage was set for a re-birth of the direction that the ancient greeks had started upon. Such was the power of the new mathematical techniques, that there emerged a new confidence in the minds of many learned people that the golden time for humanity was not necessarily back in the times of the ancients, but that the time of flourishing in the arts and science that the world experienced back in the time of the ancients could be renewed once again, that it could be renewed in the arts and sciences and in all human endeavor, and that man could once again move onward to obtain glory and great achievement. Within a century or so of the introduction of these new mathematical techniques (modern number systems and arithmetic) into Europe, the world of business, commerce, and the arts and sciences in Europe changed and along with it the new European Renaissance era began.

The subject of probability was first considered during the Renaissance primarily by Italian mathematicians. The great mathematicians Blaise Pascal and Pierre Fermat in 17th century France became very influential figures in developing the subject into a rigorous and useful form. During this time period, probability theory was developed to understand games of chance. Also about this time the first statistical ideas were being developed in England. The definition of probability that came from these early investigations relied heavily on simple counting techniques. Since then definitions of probability have changed form, but definitions based on counting techniques are still very much in use in many situations. In modern times it came to be realized that we need to use an axiomatic approach to the definition of probability, which allows for many different probability measures to be of use.

Since the European Renaissance, the world of mathematics and science has grown tremendously and we live in a truly different time, unseen in human history. The embers that remained of the ancient greek methods and ways re-ignited the scientific spirit in men, and a flood of new mathematical and scientific knowledge has overtaken modern man. We live in a new world of digital computers, modern medicine that has greatly extended the human life-span, radio and television, information that is available to people around the world on an unprecedented scale and with such great speed. We have split the atom, we have cars and airplanes, and we have the capability to reach the moon and the other planets in the solar system.

In the fundamental sciences concerned with matter, energy, and life, mainly physics, chemistry, and biology, there is still uncertainty, much that is not understood. We do not have final theories. In fact we realize that there are some things that we can never know. In mathematics itself, we have Gödels mathematical results, putting limitations on what mathematicians can prove. In physics, we have the new 20th century theories of quantum mechanics and particle physics, where probability theory is a necessary and inherent part of the theories themselves. Most modern physicists would agree with the assessment that some things are forever beyond certain knowledge, that not everything can be totally known to us, and that a certain amount of uncertainty about our world will always be with us. It seems that very possibly we will never have a "final theory."

There is a lot of research that is being done in the various branches of science. Human knowledge and technology is marching forward at an exponential pace. As stated before, uncertainty still exists and in some cases does so necessarily. In fact, the amount of uncertainty in our understanding of the world is great even though our knowledge has increased tremendously, because it seems answering a scientific question sometimes opens up many more questions. Science and mathematics will likely never lead us to a perfect understanding of the world, but scientists will continue to probe and experiment with nature because there is so much to know and so many problems to solve. Modern and truly powerful mathematical and probabilistic methods have come onto the scene only in the past 300-400 years, and most likely will remain ever important because they are so central to our scientific method.

The scientific method broadly consists of two parts, theoretical considerations and experimental verification. The theoretical part usually involves mathematical or mathematical-like reasoning, and the experimental part is the part dealing with the conducting of experiments to confirm or deny the predictions of theory. A lot of statistical techniques are used to analyze experimental results and test theoretical hypotheses. Both sides go hand in hand and work together to expand our science. Mathematical analysis is deductive. It allows us to make statements about particular cases by understanding the general. For example in Geometry we study theorems about triangles in general, from which we can make statements about particular triangles. In this way of going from the general to the specific, there is no error unless we make a mathematical error. Mathematics is the tool of reason in science. Mathematics allows us to develop theories in a clear and concise language. It is our greatest tool for understanding the world. Statistical analysis is inductive. We use statistics to make general statements based on particular results. In this way of going from the specific to the general, we run a greater risk of making errors, because particular results often cannot be generalized completely. In statistical analysis, we need to minimize the error of making false conclusions, so it is important to learn how to use statistical techniques correctly. When statistics is used correctly it is a very powerful tool. Many people that would read this book may go on to study the subject of statistics in more depth. Hopefully, the reader will gain insights from this book about mathematics and probability that they will find very helpful.

The latter part of this book explains some of the basic tools and results in probability theory, while the first part of the book explains a lot about the mathematical ideas from which these probabilistic tools and theory are derived. In the study of probability and statistics in college, I feel that undergraduates would gain more insight into probability and statistics if they were introduced to the topics of infinite number sets and the measure of numbers as presented in this book, in addition to the usual undergraduate concentration on calculus and linear algebra. The topics introduced here on infinite sets, countable and uncountable infinities, the different number systems, and also a bit about the measure of the various number sets are very important to mathematicians and statisticians, and also to physical scientists. I feel that these important topics should be introduced to students before graduate school.

CHAPTER 2:
FUNDAMENTALS OF SET THEORY

(2) FUNDAMENTAL SET NOTIONS

Set theory with its associated logic is the foundation of much of mathematical analysis. Some parts of logic are an intrinsic part of set theory. However, the subject of logic (extensive in itself and separate from set theory) will not be covered in detail in this book.

A set is a well-defined collection of objects, which we can name with a capital letter. Here in this first chapter we will mainly consider finite sets to illustrate the ideas, by which we simply mean a set with only a finite number of elements. So for example if we have a set A with three elements x, y, and z, we can write the set with its elements within braces. This is the roster method of writing a set:

$$A = \{x, y, z\}$$

Since x is a member of A, we write $x \in A$ (x is an element of A). Since w is not a member of A, we write $w \notin A$ (w is not an element of A). The order in which we list the elements is not important. So if $A = \{x, y, z\}$ and if $B = \{z, y, x\}$, then A and B are the same set. So two sets are equal if they contain exactly the same elements and we write A = B. Two sets are said to be equivalent (different from being equal) if they contain the same number of elements. So $H = \{t, g, k\}$ and $J = \{a, b, c\}$ are equivalent since they both contain three elements. We have a name for the number of elements in a set, we call it the cardinality. The cardinality of a set A we will write card(A). So the cardinality of the set $D = \{x, y, z, w, r\}$ is card(D) = 5. For any finite set D with n elements in it, card(D) = n. The null set or the empty set, the set with no elements in it, is written $\oslash$, and the cardinality of the null set is zero. Therefore, for any finite set with cardinality n, n is an integer greater than or equal to zero. For infinite sets, the cardinality is some type of so-called transfinite number. We'll discuss this in more detail in the next chapter where we will consider infinite sets. As another example, consider the set

$$C = \{a, b, c, d, e, f, g, h, i, j, k, l, m, n, o, p, q, r, s, t, u, v, w, x, y, z\} .$$

These are the 26 lowercase letters of the English alphabet, written in roster form. The card(C) = 26. Another important way of writing a set is the so-called property method. Doing this for the set C, we re-write the description of C as:

$$C = \{x \mid x \text{ is a lowercase letter of the english alphabet}\}$$

This is stated: C is the set that contains elements x, where x is a lowercase letter of the English alphabet. This method is important for sets that contain a large number of elements, where we can't write out all of its elements in roster form. Many times this would be necessary for sets of numbers. For example, we may have the set

$$E = \{x \mid x \text{ is an integer, } x \geq 4, \, x \neq 10\}$$

E is the set of all integers which are greater than or equal to 4, but excluding 10. The cardinality of E is a topic that we will leave to the next chapter. Another way of writing E would be:

$$E = \{4, 5, 6, 7, 8, 9, 11, 12, 13, \ldots\}$$

This is just another form of the roster method. But note that there is a problem in this particular case, since by excluding 10 we may not know whether or not we mean to exclude certain other numbers in this set, such as every 7th term or so. Obviously it is not best to try to write this set E in roster form, the property method is definitely best. However, in other cases there may be no ambiguity, such as for the following set G:

G = $\{0, 1, 0, 1, 0, 1, \ldots\}$

We can be reasonably sure of what's going on with the set G. Still another way to specify a set is to have a special symbol for the set that is universally understood, such as $\mathbb{R}$ for the real numbers.

So, how we write a set is usually decided for a particular situation, but the roster method, the property method, or providing a special letter to denote a set are the most common ways. There is another way that many real number sets can be written, which is called interval notation. For example a set written as (a,b), where a < b, is the set of real numbers that are between a and b but not including a or b. This is an open set. A set written as [a,b], where a < b, is the set of real numbers that are between a and b and including a and b. This is a closed set. The half open/half closed set (a,b], where a < b, is the set of real numbers that are between a and b, not including a but including b. The half open/half closed set [a,b), where a < b, is the set of real numbers that are between a and b, not including b but including a. When an interval extends to -∞, it is preceded by a parenthesis. When an interval extends to ∞, it is followed by a parenthesis. The real numbers $\mathbb{R} = (-\infty, \infty)$.

(3) <u>SUBSETS, UNIONS, INTERSECTIONS</u>

Set A is a subset of a set B if all of the elements of A are contained within set B. For example, if B = $\{a, b, c, d, e, f, g, h\}$, the following sets $A_1 = \{a, d, e\}$, $A_2 = \{a, b, c, d, e, f\}$, and $A_3 = \{g\}$ are all subsets of B. They are proper subsets since they do not contain all of the elements in B. Any Set C obviously contains all of the members in itself. In this case we would also say that C is a subset of itself. So any set is a subset of itself. The null set $\oslash$ is a subset of C, in fact $\oslash$ is a subset of all sets including itself. When a set A is a subset of a set C we write A $\subset$ C.

The union of two sets A and B is the set which contains all elements that are in A or in B. For example, if A = $\{1, 2, 5, 6\}$ and B = $\{3, 4, 5, 6, 7, 8, 9, 10\}$, then the union of A and B, written A $\cup$ B, is A $\cup$ B = $\{1, 2, 3, 4, 5, 6, 7, 8, 9, 10\}$. It is obvious that for any set A, A $\cup$ A = A and A $\cup$ $\oslash$ = A.

The intersection of two sets A and B is the set which contains all elements that are in A and in B. For the sets A and B of the previous paragraph, the intersection of sets A and B, written A $\cap$ B = $\{5, 6\}$ since these are the only elements that are common to both sets. It is obvious that for any set A, A $\cap$ A = A and A $\cap$ $\oslash$ = $\oslash$.

(4) SET COMPLEMENTS

If we have a universal set Ω (which contains all elements in some space of interest) and a set A that is a subset of Ω, then the complement of A relative to Ω is the set containing all elements in Ω that are not in A, and we write it $\widehat{A}$. For example if the universal set $\Omega = \{0, 1, 2, 3, 4, 5, 6, 7, 8, 9\}$ and if A = $\{0, 5, 9\}$, then $\widehat{A}$ = $\{1, 2, 3, 4, 6, 7, 8\}$. It is obvious that $A \cup \widehat{A} = \Omega$ and $A \cap \widehat{A} = \oslash$. It's also true that $\widehat{\oslash} = \Omega$ and $\widehat{\Omega} = \oslash$. For any set A, A and $\widehat{A}$ are said to be disjoint or mutually exclusive since they have no elements in common. More generally, any two sets that have no elements in common are said to be disjoint or mutually exclusive, and if we have a collection of sets $\{A_1, A_2, \ldots, A_n\}$

such that $\bigcup_{i=1}^{n} A_i = \Omega$ and all of the sets are pairwise disjoint, then we say that we have a mutually exclusive and exhaustive collection of sets. A result that is often important in set theory is that for any set E and any set A, E is the part of E that is in A plus the part of E that is in $\widehat{A}$. Therefore we have $E = (E \cap A) \cup (E \cap \widehat{A})$.

(5) <u>THE ALGEBRA OF SETS</u>

With real numbers, we have an algebra defined for the real numbers including addition and subtraction, multiplication and division (with all the commutative, associative, and distributive laws associated with these operations), along with relations like equality and orderings. From the discussion of this chapter we can see that there is an algebra for collections of sets as well. This set algebra consists of the operations of unions, intersections, complements (along with the corresponding commutative, associative and distributive laws associated with these operations), along with relations like equality and inclusion (one set being a subset of another).

(6) <u>RELATIONS AND FUNCTIONS</u>

A relation (in this book) is any collection (or set) of points in the two-dimensional coordinate plane. An example of a relation is the set of points R = { (1,1), (-2,5), (-1,10), (-1,2), (-1,-4), (7,3), (2,3), (3,2) }. The point (2,3) means that 2 is related to 3, the point (-2,5) means that -2 is related to 5, and so on. Note that for this relation, the real number -1 is mapped, or related, to three distinct real numbers. Contrast this below with functions, where there can only be one y-value for each x-value. A familiar example of a relation is a circle; and we know that it is an infinite collection of points in the plane. If we have any vertical line in the plane, a relation allows for that vertical line to intersect the relation in more than one point. For a function of one variable x, which is also a finite or infinite collection of points in the plane, any vertical line must intersect the function in only one point. A circle, which is a relation with an infinite number of points, is a good example of a relation because some vertical lines intersect the circle in two points. We see that a function is a special type of relation. To analyze a circle in terms of functions, we have to consider the top half and the bottom half of the circle as being two separate functions.

To introduce the concepts associated with functions we will use finite discrete functions, that is, functions that consist of a finite collection of points. Continuous functions of a real variable will be discussed considerably later on in the book when we go over the important points of differential and integral calculus. A function f(x) has associated with it a set X called the domain, a set Y called the co-domain, and a set f(X) called the range. The function f(x) is a rule that assigns only one y-value in the co-domain Y to every x-value in the domain X. So we sometimes write $f : X \rightarrow Y$ (f maps X to Y). The range f(X) is the set of all y-values such that there is at least one x-value (it's pre-image) where y = f(x). Therefore the range f(X) is a subset of Y.

In most applications of math involving functions, y = f(x) is given as an explicit formula such as: $y = f(x) = x^3 + 13x - 9$, $y = f(x) = 1 + \sin(4x)$, or $y = f(x) = xe^x$, and so on. Functions like these usually have domain and co-domain consisting of some infinite subset of the real numbers $\mathbb{R}$. We see functions like these frequently in calculus.

<u>Example 1</u>:   To illustrate the ideas of functions being into, onto, one-to-one, and having an inverse function, we will use as a first

example the set $\{1,2,3,4,5,6,7,8\}$ for both the domain X and co-domain Y. Here I'm using the same set with the same cardinality for X and Y. There is no particular reason for this. There are many different choices for the set X and Y that can be used to illustrate the ideas. Let f(x) be the function defined as:

f(x) = { f(1) = 1, f(2) =2, f(3) =2, f(4) = 4,
    f(5) = 4, f(6) = 7, f(7) = 7, f(8) = 7 }.

This is a function which maps X into Y, since f(X) ={1,2,4,7} is a subset of Y. There are elements in Y which do not have a pre-image in X, so it is not onto. An onto function is one where each element of the co-domain Y has a pre-image x in the domain such that y = f(x). A function is one-to-one if for any two distinct numbers a and b in the domain X, then f(a) ≠ f(b). Note that f(x) is not a one-to-one function between X and Y, or between X and f(X), so there is no inverse function (or reverse one-to-one and onto mapping) from Y to X or from f(X) to X.

As another example, let f(x) be the function defined as:

f(x) = { f(1) = 1, f(2) =8, f(3) = 2, f(4) = 7,
    f(5) = 3, f(6) = 6, f(7) = 4, f(8) = 5 }.

This function is into Y and also maps X onto Y since f(X) = {1,2,3,4,5,6,7,8} = Y. This is also a one-to-one function since if we have two different x-values a and b, then f(a) ≠ f(b). It pairs up the numbers in the sets X and Y in a one-to-one way. When we have a one-to-one and onto function from X to Y such as this one, then we have a unique inverse function $f^{-1}(y) : Y \rightarrow X$ such that f⁻¹(f(x)) = x for every x in X, and f(f⁻¹(y)) = y for every y in Y. In this case:

f⁻¹(y) = { f⁻¹(1) = 1, f⁻¹(2) = 3, f⁻¹(3) = 5, f⁻¹(4) = 7,

$f^{-1}(5) = 8$, $f^{-1}(6) = 6$, $f^{-1}(7) = 4$, $f^{-1}(8) = 2$ }.

Clearly there are many different into, onto, one-to-one, and inverse functions that can be defined here with the chosen sets X and Y, and in addition, we could have chosen many different sets X and Y to be the domain and co-domain. So let's consider another example.

Example 2: Let X = {a,b,c,d,e,s} and Y = {1,2,3,4} be the domain and co-domain, respectively. The function f(x) is defined:

$f(x) = \{ f(a) = 3, f(b) = 3, f(c) = 3, f(d) = 3, f(e) = 3, f(s) = 3 \}$.

This is called a constant function since the y-value is the same for every x-value in the domain. It's an into function with range f(X) = {3}. But we can see that it's not a function onto Y, neither is it one-to-one. So we don't have the possibility of an inverse function from Y to X or from f(X) to X. Now consider the function f(x):

$f(x) = \{ f(a) = 1, f(b) = 1, f(c) = 2, f(d) = 3, f(e) = 4, f(s) = 4 \}$.

This function is into Y and onto Y, so the range f(X) = Y. However it is not one-to-one since both a and b are mapped to 1, and both e and s are mapped to 4. So we don't have an inverse function from Y to X. As another example, let f(x) be defined:

$f(x) = \{ f(a) = 1, f(b) = 2, f(c) = 3, f(d) = 4, f(e) = 4, f(s) = 4 \}$.

This function is into and onto, but it is not a one-to-one function from X to Y. So we can't have an inverse function from Y to X. However, if we restrict the domain X to the set B = {a,b,c,d}, then the resulting function f(x) from B $\rightarrow$ f(B) defined as:

$f(x) = \{ f(a) = 1, f(b) = 2, f(c) = 3, f(d) = 4 \}$

is a one-to-one and onto function from B → f(B). Therefore we can define an inverse function f^{-1} from f(B) → B like so:

$f^{-1}(y) = \{ f^{-1}(1) = a, f^{-1}(2) = b, f^{-1}(3) = c, f^{-1}(4) = d \}$.

Restricting the domain to get a useful function is common in mathematics. One case that comes to mind is with the inverse trigonometric functions. For example, consider the tangent function, where we restrict the domain of the tangent function to $(-\pi/2, \pi/2)$. Then the tangent function is a one-to-one function from this interval onto the set of real numbers, so we can define the arctan function (inverse tangent function) from the real numbers to the set $(-\pi/2, \pi/2)$.

(7) **SOME COMBINATORICS**

Because we use counting techniques for finite sets, especially in one of the definitions of probability that we will discuss later in the book, we will consider some elementary topics from the mathematical subject of combinatorics. These topics allow us to figure out the number of things in certain sets, and the number of ways that certain things can occur. Combinatorics is a very extensive subject within mathematics. For our purposes in this book, we will consider only a few important techniques.

The Fundamental Principle of Counting

In combinatorics, we have a starting point that we call the Fundamental Principle of Counting (FPC). It says that if event A can happen in 'n' ways and event B can independently occur in 'm' ways, then events A and B can happen together in (n · m) ways. This can obviously be extended to more than two events. If events $A_1, A_2, \ldots, A_r$ can independently occur in $n_1, n_2, \ldots, n_r$ ways respectively, then they can all happen together in $(n_1) \cdot (n_2) \cdots (n_r)$ different ways.

Permutations and Combinations

If we have n objects, say for example lined up in a row, different orderings are called different permutations of those n objects. One may ask how many different permutations of those n objects are there? If we imagine n locations where the objects can be placed, the first place can be filled in (n) ways, after that the second place can be filled in (n - 1) ways, after that the third place can be filled in $(n-2)$ ways, and so on until we get to the nth location which can obviously be filled in only (1) way. The product of these ways is n! = (n)(n - 1)(n - 2)$\cdots$(2)(1) from the FPC. So the number of different permutations of n objects is n!, pronounced "n factorial." If we have n objects and only r locations $(r < n)$, then the number of permutations of r things chosen from n things would by an analogous argument be $(n)(n-1)(n-2)\cdots(n-r+1)$. This is written P(n,r).

$$P(n,r) = (n)(n-1)(n-2) \cdots (n-r+1) = \frac{(n)(n-1)(n-2)\cdots(3)(2)(1)}{(n-r)(n-r-1)\cdots(3)(2)(1)} = \frac{n!}{(n-r)!} \; .$$

P(n,r) is the total number of ways that we could select r things from n things and arrange them in some order. The order of the r things once chosen from the n things is important when talking about permutations.

If we are only interested in the total number of distinct collections of r things that we could choose from n things, without regard to order, then we simply divide P(n,r) by (r!). This is written C(n,r) and called the number of combinations of r things chosen from n.

$$C(n,r) = \frac{P(n,r)}{r!} = \frac{n!}{(n-r)!(r)!} \; .$$

C(n,r) is the total number of subsets of size r from a total of n things, and as we know to be true of sets the order of the r things is not important. With some thought one can see that the total number of subsets of n things is C(n,0) + C(n,1) + $\cdots$ + C(n,n). This turns out to be 2^n. C(n,r) is also the number of ways that we could partition n things into two groups of size (r) and (n - r). C(n,r) is often pronounced "n choose r," and is also called a binomial coefficient because of its association with the binomial distribution, something we will study later.

Partitions

If we have n things and want to know the number of ways that we could partition the n things into r groups of sizes $n_1, n_2, \ldots, n_r$ where $n_1 + n_2 + \cdots + n_r = n$, then we could do it this way: First select n_1 things from the n things, which could be done in $C(n, n_1)$ ways. Then select n_2 things from the remaining (n - n_1) things, which could be done in $C((n - n_1), n_2)$ ways, and so on. Denote the number of partitions of n things into r groups of sizes $n_1, n_2, \ldots, n_r$ with Part(n|$n_1, n_2, \ldots, n_r$). Then:

Part(n|$n_1, n_2, \ldots, n_r$) = C(n,n_1) $\cdot$ C((n - n_1),n_2) $\cdots$ C((n - n_1 - n_2 - $\cdots$ - n_{r-1}),n_r)
which turns out to be

$$\frac{(n!)}{(n_1!)(n_2!)\cdots(n_r!)}$$

Note that C(n,r) is a special case of the partition approach. We could write C(n,r) = Part(n|r,n - r).

As an example, how many partitions of 7 objects into three groups of size 2, 2, and 3 are there? First figure the number of ways that 2 things could be chosen from the 7, C(7,2). Then figure how many ways 2 things could be chosen from the remaining 5, C(5,2), and finally how many ways can 3 things be chosen from the remaining 3, C(3,3).

So Part(7|2,2,3) = C(7,2) · C(5,2) · C(3,3) =

$$\frac{7!}{(5!)(2!)} \cdot \frac{5!}{(3!)(2!)} \cdot \frac{3!}{(0!)(3!)} = \frac{7!}{(2!)(2!)(3!)} = 210. \quad \text{(we define 0! = 1).}$$

CHAPTER 3:
FUNDAMENTALS OF INFINITE SETS

(8) <u>**NUMBER SETS**</u>

Infinite sets are the sets usually of most interest and difficulty in set theory. This generally means various types of number sets, such as the Natural numbers, Whole numbers, Integers, Rational numbers, Irrational numbers, the Real numbers, and finally the Complex numbers. The starting point is the Natural numbers or Counting numbers as they are sometimes called. We give them the symbol $\mathbb{N}$:

$$\mathbb{N} = \{1,2,3,4,5, \ldots\} \, .$$

This is undoubtedly the numbers that humans first used. The development of zero was slow, but Asian and Arabic scholars of the middle ages introduced the concept into their number systems and developed our modern positional number system along with all the associated arithmetic techniques. The Natural numbers along with zero are called the whole numbers $\mathbb{W}$:

$$\mathbb{W} = \{0,1,2,3,4,5, \ldots\} \, .$$

The next great development was the introduction and acceptance of the idea of negative numbers. So we had the negative and positive natural numbers, and the number 0. Collectively, we call these the integers $\mathbb{Z}$:

$$\mathbb{Z} = \{\ldots,-4,-3,-2,-1,0,1,2,3,4, \ldots\} \, .$$

The next stage in the development of our modern number system was the inclusion of fractions. Taking the integers and combining them with all the ratios of integers (without dividing by zero) led to the important set of numbers that we call the rational numbers $\mathbb{Q}$:

$\mathbb{Q} = \{x|$ x is a number of the form $(\frac{a}{b})$, where a, b $\in \mathbb{Z}$, b $\neq 0$ $\}$.

The greek mathematicians thought that all numbers were rational. When they proved that the diagonal length of a square with sides equal to 1 was not rational, their cherished notions about numbers and geometry were turned upside down. In modern times, we simply say that if a real number is not rational, then it is automatically irrational. No other alternative type of number is known to exist in order to complete the real numbers. Without the irrational numbers, the number line is incomplete. It has holes in it so to speak.

So we arrive at the real numbers, given the symbol $\mathbb{R}$, which we believe to be simply the union of the disjoint sets of rational and irrational numbers. The real numbers constitute what is believed to be a continuum: for every real number there is a unique point on the number line and for every point on the number line there is a unique real number. We refer to this as the real line in one dimension. The real numbers are what is called "ordered," meaning that if x and y are any two real numbers, then either x < y, x = y, or x > y. Whenever we set up a Cartesian coordinate system in two or three dimensions, we assume that the axes are real lines. A real number is all possible decimal expansions (base 10), or all possible binary expansions (base 2), or all possible hexadecimal expansions (base 16), or all possible expansions with whatever base one chooses. Most commonly we use decimal expansions in our modern mathematics (aside from much of the mathematical theory and techniques associated with computers, and similar fields of study). It is the case that all rational expansions must repeat a pattern after a certain point, but all irrational number expansions go on indefinitely without ever repeating any pattern, no matter what base you use.

There is another subdivision of the real numbers which is often of interest to mathematicians. The set of all real numbers that are roots of some nth degree polynomial with rational coefficients are called the algebraic numbers. This set contains all the rationals and an infinite number of irrationals as well, but it turns out that the algebraic numbers are countably infinite. (Countable infinity will be discussed soon. We will show that the rationals are countably infinite, but we will not show that the algebraic numbers are countably infinite. To demonstrate this is not terribly difficult, but it is time consuming and tedious. We will not do that here). All the remaining real numbers are known as the transcendental numbers, all of them irrationals, and it turns out that they are uncountably infinite (Once again this is to be discussed in a later section). It is easy to write down examples of algebraic numbers, but aside from e (the base of the natural exponential function and the natural logarithmic function) and π, it isn't easy to think of examples of transcendental numbers. Yet the transcendental numbers constitute a higher infinity of numbers. So the real numbers are equivalent to a subset of the irrationals!

Finally we have the Complex numbers $\mathbb{C}$. The complex numbers are a two dimensional number, To each point of the coordinate plane (a,b) there corresponds the complex number a + bi, where a and b are real numbers (a is called the real part and b is called the imaginary part) and $i = \sqrt{-1}$. The real numbers are just the numbers on the horizontal axis, and the purely imaginary numbers are just the numbers on the vertical axis. All other numbers in the so-called complex plane contain a non-zero real and a non-zero imaginary component. One of the most important properties of the complex numbers is that they are what is called algebraically complete. This means that any nth degree polynomial with real coefficients will always have exactly n complex roots (some of them are often real), and therefore the polynomial can then be factored into n linear factors.

(9) <u>COUNTABLE INFINITY</u>

The famous mathematician George Cantor is responsible, just about single handedly, for developing our ideas about sets, infinite sets, and the various types of infinity. He noticed that when we count n objects, we associate 1 with the first object, 2 with the second object, and so on until we associate the integer n with the last object. It is by a 1-1 correspondence with the positive integers like this that we count a finite collection of objects. He extended this idea to infinite sets. The most fundamental type of infinity is that of the positive integers, or natural numbers $\mathbb{N} = \{1,2,3, \ldots\}$. Any set that can be put into 1-1 correspondence with the positive integers is the smallest type of infinity, which we call countable infinity. The symbol that we will use for this type of infinity is $\aleph_0$. This is the cardinality of this smallest type of infinity, it is called a transfinite number. Note that the sets $\mathbb{N}$ and $A = \{2,4,6, \ldots\}$ can be put into a 1-1 correspondence using the function f(n) = 2n. So they are said to contain the same number of objects, hence the cardinality of A is $\aleph_0$. We say that sets A and $\mathbb{N}$ are equivalent, both have cardinality $\aleph_0$, but $A \subset \mathbb{N}$. This is a property of infinite sets: they are equivalent to a proper subset of themselves. Similarly B = {10,20, . . .} can be put into 1-1 correspondence with $\mathbb{N}$ using the function f(n) = 10n. So clearly the cardinality of B is $\aleph_0$ as well. One can see that there are an infinite number of sets with cardinality $\aleph_0$.

More difficult to see at first, yet true, is that the set of rational numbers $\mathbb{Q}$ is also countably infinite and has cardinality $\aleph_0$. We will show that this is true. Consider the following doubly infinite array of fractions. All possible rational numbers (ratios of integers without dividing by zero) are included in the array and of course many represent the same number. We could eliminate repeats as we go along, but it is not necessary to the argument. Follow the indicated path shown by the arrows, starting with $\left(\frac{0}{1}\right)$ in the upper left:

$$\left(\tfrac{0}{1}\right) \Rightarrow \left(\tfrac{0}{-1}\right) \qquad \left(\tfrac{0}{2}\right) \Rightarrow \left(\tfrac{0}{-2}\right) \qquad \left(\tfrac{0}{3}\right) \Rightarrow \left(\tfrac{0}{-3}\right) \cdots$$
$$\swarrow \qquad \nearrow \qquad \swarrow \qquad \nearrow \qquad \swarrow$$
$$\left(\tfrac{1}{1}\right) \quad \left(\tfrac{1}{-1}\right) \quad \left(\tfrac{1}{2}\right) \quad \left(\tfrac{1}{-2}\right) \quad \left(\tfrac{1}{3}\right) \quad \left(\tfrac{1}{-3}\right) \cdots$$
$$\Downarrow \quad \nearrow \qquad \swarrow \qquad \nearrow \qquad \swarrow$$
$$\left(\tfrac{-1}{1}\right) \quad \left(\tfrac{-1}{-1}\right) \quad \left(\tfrac{-1}{2}\right) \quad \left(\tfrac{-1}{-2}\right) \quad \left(\tfrac{-1}{3}\right) \quad \left(\tfrac{-1}{-3}\right) \cdots$$
$$\swarrow \qquad \nearrow \qquad \swarrow$$
$$\left(\tfrac{2}{1}\right) \quad \left(\tfrac{2}{-1}\right) \quad \left(\tfrac{2}{2}\right) \quad \left(\tfrac{2}{-2}\right) \quad \left(\tfrac{2}{3}\right) \quad \left(\tfrac{2}{-3}\right) \cdots$$
$$\Downarrow \quad \nearrow \qquad \swarrow$$
$$\left(\tfrac{-2}{1}\right) \quad \left(\tfrac{-2}{-1}\right) \quad \left(\tfrac{-2}{2}\right) \quad \left(\tfrac{-2}{-2}\right) \quad \left(\tfrac{-2}{3}\right) \quad \left(\tfrac{-2}{-3}\right) \cdots$$

By following the path shown in the array we see how the set of all fractions, the rationals $\mathbb{Q}$, can be put into a 1-1 correspondence with the positive integers $\mathbb{N}$. Therefore the rationals are countably infinite and have cardinality $\aleph_0$. Note that we can write in a roster form the elements of a countably infinite set. For the rationals (following the path above through the two-dimensional infinite matrix):

$$\mathbb{Q} = \left\{ \left(\tfrac{0}{1}\right), \left(\tfrac{0}{-1}\right), \left(\tfrac{1}{1}\right), \left(\tfrac{-1}{1}\right), \left(\tfrac{1}{-1}\right), \left(\tfrac{0}{2}\right), \left(\tfrac{0}{-2}\right), \left(\tfrac{1}{2}\right), \left(\tfrac{-1}{-1}\right), \cdots \right\}$$

(10) UNCOUNTABLE INFINITY

We now wish to show that there is a higher level of infinity than countable infinity. We call this uncountable infinity. Mathematicians have spent many years trying to find a level of infinity between the two, but their efforts so far have ended in failure. The sets that we are familiar with that are uncountably infinite are the irrationals, the real numbers, and the complex numbers. We will demonstrate a way that mathematicians have used to prove the existence of uncountable infinity. We will start with the real number interval [0,1] and we will show that this interval is uncountably infinite. Let's assume that the real numbers in [0,1] are countable and show that this leads to a contradiction. If they are countable, then they can be exhibited in a list of decimal expansions. We'll let $(0 = 0.0000...)$ and $(1 = 0.9999...)$. All subscripted letters are one of the ten digits 0 through 9. Our list of the reals in [0,1] is:

$$0. \, a_1 a_2 a_3 a_4 \, . \, . \, . \, .$$
$$0. \, b_1 b_2 b_3 b_4 \, . \, . \, . \, .$$
$$0. \, c_1 c_2 c_3 c_4 \, . \, . \, . \, .$$
$$0. \, d_1 d_2 d_3 d_4 \, . \, . \, . \, .$$

Now let's construct a new number $z = (0.z_1 z_2 z_3 z_4 \, . \, . \, . \, .)$, where $z_1 \neq a_1$, $z_2 \neq b_2$, $z_3 \neq c_3$, $z_4 \neq d_4$, and so on. We can see that z differs from the first number in the list in at least the first decimal place, z differs from the second number in the list in at least the second decimal place, z differs from the third number in the list in at least the third decimal place, and so on. This means that z (which is a real number in [0,1]) is not in our list, contrary to our assumption that all real numbers in [0,1] are somewhere in the list because they are countable. We have arrived at a contradiction that all real numbers in [0,1] are countable and can be listed. Since we could just add z to the list and continue with the argument above as much as we want, we have to conclude that there is a higher level of infinity, which we call uncountable infinity. The cardinality of an uncountably infinite set is denoted $\aleph_1$.

Define a 1-1, onto function f(x) from (0,1) to $(-\infty, \infty)$ as
$f(x) = \tan(-\frac{\pi}{2} + \pi x)$, for all x in (0,1). Since we know (0,1) is uncountably infinite, then the real numbers $\mathbb{R}$ are also uncountably infinite, because the function establishes a 1-1 and onto correspondence between (0,1) and $(-\infty, \infty)$. It is the case that removing the countable set of rationals from the real numbers will leave behind the uncountably infinite set of irrationals. The real numbers are equivalent to a proper subset of themselves, the irrationals. We stated above that this is a property of infinite sets.

At this point we will state that the rationals are so-called dense in the real numbers. This means that every real number is either a rational number or a limit point of the rationals. A limit point z of a set A of real numbers is a real number z such that there is a sequence $\{x_1, x_2, x_3,...\}$ of numbers from A that converges to z. We will talk more about convergence in chapter 4 when we talk about convergent sequences. However, the reader probably knows what is meant. As an illustrative example, Let A be the set of all non-zero rationals in [0,1] and consider the sequence $\{\frac{1}{2}, \frac{1}{4}, \frac{1}{8}, \frac{1}{16}, \frac{1}{32}, \frac{1}{64},...\}$. This set of numbers converges to the real number 0, but 0 is not in the set A. So this set A doesn't contain its limit point 0. One can have a sequence of rationals that converges to $\sqrt{2}$, but obviously $\sqrt{2}$ is not rational. The Rational numbers don't contain all of their limit points, but the real numbers do. The set of real numbers are so-called complete since they contain all of their limit points.

(11) <u>HIGHER ORDERS OF INFINITY</u>

It turns out that we can find an infinite hierarchy of infinite sets with cardinalities $\aleph_2$, $\aleph_3$, $\aleph_4$, ... However they are associated with ever more complicated geometric sets than the real numbers. For example the set usually used to demonstrate an infinite set with cardinality $\aleph_2$ is the set of all curves in the two-dimensional Cartesian coordinate plane, which we will not consider in detail. So in this book on real analysis we only have to deal with countably and uncountably infinite sets.

CHAPTER 4:
FUNDAMENTALS OF CALCULUS

(12) **THE CALCULUS**

When we study high school geometry, we deal with geometric figures mostly constructed from straight lines and circles. For example triangles, trapezoids, parallelograms, angles, arcs and so on. For these simple figures, we can calculate lengths, areas, volumes, and so on with relatively easy mathematics. But we know that the real world involves a lot of geometric objects that have many other kinds of shapes. Curves, surfaces, and volumes based on continuous functions in the plane and 3-dimensional space often involve all kinds of irregular shapes. Calculus enables us to analyze more complicated geometric and more complicated algebraic problems in the various sciences and engineering. At the root of the more complicated tools that we have in calculus is the idea of the "rate of change" of continuously varying quantities with respect to a change in another quantity. For example for a function $y = f(x)$, we may be interested in the instantaneous rate of change of y with respect to a change in x at a point on a curve. Therefore, we can think of calculus as the mathematics of "change." The main tool that we start with is called the derivative. From the derivative we develop another tool known as the definite integral. Both the derivative and the definite integral allow us to analyze those more complicated problems that arise in the real world. Let me start with the ideas of an infinite sequence and an infinite series.

(13) **SEQUENCES**

A sequence is a function that has the positive integers as its domain. It can also be thought of as an ordered, countable, set of numbers like $\{\frac{1}{2}, \frac{1}{3}, \frac{1}{4}, ...\}$. We say ordered since each number has its own place in the sequence. In other words, unlike a more general set where the order in which the elements are listed is unimportant, for a sequence there is a definite order in the listing of the elements, which is important. Many times we will write a sequence more abstractly, such as $\{x_1, x_2, x_3, ...\}$ or simply as $\{x_n\}$. A sequence either converges or diverges. Convergent means that it gets arbitrarily close to a single number for all terms after some point. A more technical definition of convergence is:

A sequence $\{x_n\}$ converges to a real number L, if for any $\varepsilon > 0$ (ε is a positive real number that we can choose to be as small as we like), there exists a positive integer N such that $|x_n - L| < \varepsilon$, whenever $n > N$.

An alternate definition of convergence, which is equivalent to the first, and called Cauchy convergence, says:

A sequence $\{x_n\}$ converges to some real number L, if for any $\varepsilon > 0$, there exists a positive integer N such that $|x_n - x_m| < \varepsilon$, whenever n and m are greater than N.

A divergent sequence means that the terms don't continue to get arbitrarily close to some number L for all terms after some point. For example the sequence $\{1,0,1,0, . . .\}$ never gets arbitrarily close to 0 or 1, or any other number L, so it is divergent. However, this sequence has two convergent subsequences, $\{1,1,1, . . .\}$ and $\{0,0,0,...\}$, converging to 1 and 0 respectively. It should be obvious that if a sequence converges to a number L, then all of its subsequences converge to L also, and vice versa. As another example the sequence $\{1,2,3,4, . . .\}$ diverges because the terms are going to ∞ , instead of a single number L.

Many times if a sequence $\{x_n\}$ converges to a number L, we just simply write $\lim_{n\to\infty} (x_i) = L$, or $\{x_n\} \to L$.

(14) <u>INFINITE SERIES</u>

An infinite series is an infinite sum like $\sum\limits_{n=1}^{\infty} a_n$ or $\sum\limits_{n=0}^{\infty} a_n$ and so forth, where each term depends on n. A series can converge (add up to a finite number) or diverge (add up to infinity, or the sum can oscillate and fail to approach a specific number). If a series converges, then the $\lim\limits_{n\to\infty} a_n$ must be zero, but the converse is not necessarily true.

The best way to analyze the convergence or divergence of an infinite series is to investigate the sequence of partial sums. For $\sum\limits_{n=1}^{\infty} a_n$, define:

$$S_1 = a_1, \quad S_2 = a_1 + a_2, \quad S_3 = a_1 + a_2 + a_3, \; \ldots$$

$\{S_n\}$ is the sequence of partial sums. If $\{S_n\}$ converges, then the series converges. If $\{S_n\}$ diverges, then the series diverges.

<u>Example 1</u>: $\sum\limits_{n=1}^{\infty} \left(\frac{1}{n^2}\right) = \left(1 + \frac{1}{4} + \frac{1}{9} + \frac{1}{16} + \frac{1}{25} + \cdots\right)$ has partial sums

$\{S_n\} = \{1, \; \frac{5}{4}, \; \frac{49}{36}, \; \frac{205}{144}, \; \frac{5269}{3600}, \; \cdots\}$

$\quad = \{1, \, 1.25, \, 1.3611, \, 1.4236, \, 1.4636, \, \cdots\}$

which can be shown to converge in the following way:

Let the function
$g(x) = \{ \; 1 \;\; \text{for } 0 \le x \le 1, \; \frac{1}{4} \;\text{for } 1 < x \le 2, \; \frac{1}{9} \;\text{for } 2 < x \le 3, \; \cdot \;\cdot\;\cdot \}$

Then $\sum\limits_{n=1}^{\infty} (\frac{1}{n^2}) = \int\limits_{0}^{\infty} g(x)dx < 1 + \int\limits_{1}^{\infty} (\frac{1}{x^2})dx = 2$, because the curve

$y = \frac{1}{x^2}$ is above g(x) from x = 1 onward to the right, and the integral = 1.

Therefore, $\sum\limits_{n=1}^{\infty} (\frac{1}{n^2})$ converges to some number < 2. The sequence of

partial sums {S$_n$} is monotonically increasing and bounded above, so it will converge to some number, in this case some number less than 2.

Many times we are not interested in what a series will converge to, but rather just whether or not it converges. If we can show that it converges then we can get a value for the sum to any degree of accuracy by using a computer to evaluate the sum for a finite number of terms.

Example 2: The $\sum\limits_{n=1}^{\infty} (\frac{1}{n}) = (1 + \frac{1}{2} + \frac{1}{3} + \frac{1}{4} + \frac{1}{5} + \cdots)$ has partial sums

{S$_n$} = {1, $\frac{3}{2}$, $\frac{11}{6}$, $\frac{25}{12}$, $\frac{137}{60}$, $\cdots$}
    = {1, 1.5, 1.8333, 2.0833, 2.2833, $\cdots$}

which can be shown to diverge in the following way.

Let the function
g(x) = {1 for $1 \le x \le 2$, $\frac{1}{2}$ for $2 < x \le 3$, $\frac{1}{3}$ for $3 < x \le 4$, ...}

Then $\sum\limits_{n=1}^{\infty} (\frac{1}{n}) = \int\limits_{1}^{\infty} g(x)dx > \int\limits_{1}^{\infty} (\frac{1}{x})dx = \infty$, because the curve $y = \frac{1}{x}$

is below g(x) for x = 1 onward to the right, and the integral = ∞.

Therefore, $\sum\limits_{n=1}^{\infty} (\frac{1}{n})$ diverges. We call this the harmonic series.

Example 3: $\sum\limits_{n=0}^{\infty} (-1)^n = (1 - 1 + 1 - 1 + 1 - 1 + \cdot \cdot \cdot)$ has partial sums

{S$_n$} = {1, 0, 1, 0, 1, 0, $\cdot \cdot \cdot$} which does not converge to a single number, so we say that this series diverges.

<u>Example 4</u>: $\sum\limits_{n=1}^{\infty} (-1)^{n+1}\left(\frac{1}{n}\right) = (1 - \frac{1}{2} + \frac{1}{3} - \frac{1}{4} + \frac{1}{5} - \frac{1}{6} + \cdot\cdot\cdot)$ has

partial sums $\{S_n\} = \{1, \frac{1}{2}, \frac{5}{6}, \frac{7}{12}, \frac{47}{60}, \frac{37}{60}, \cdot\cdot\cdot\}$

$\qquad\qquad = \{1, 0.5, 0.8333, 0.5833, 0.7833, 0.6166, \cdot\cdot\cdot\}$

which converges because the difference between the sum of the series L (whatever it is) and the first n terms is no more than $\frac{1}{n+1}$. So the

$\lim\limits_{n\to\infty} |L - S_n| \le \lim\limits_{n\to\infty} \frac{1}{n+1} = 0$. This is called an alternating series and

converges if $\lim\limits_{n\to\infty} a_n = 0$ and $|a_{n+1}| \le |a_n|$, (for all n = 1,2,3,...) .

A series is said to be absolutely convergent if $\sum\limits_{n=1}^{\infty} |a_n|$ converges. This

is the strongest type of convergence for an infinite series. If

$\sum\limits_{n=1}^{\infty} a_n$ is absolutely convergent, then the series will still converge if we

insert parentheses around collections of terms here and there, or make some of the terms negative here and there, and so on. It is called

conditionally convergent if $\sum\limits_{n=1}^{\infty} a_n$ converges, but $\sum\limits_{n=1}^{\infty} |a_n|$ diverges. The

series of example 1 is absolutely convergent. The series of example 4 is conditionally convergent but not absolutely convergent, because if we make all the terms positive we would have the divergent harmonic series.

(15) <u>GEOMETRIC SERIES</u>

A particularly important type of infinite series is the geometric series,

which is of the form $\sum\limits_{n=0}^{\infty} (a)\,(r)^n$, where a is any real number and r is a

real number with |r| < 1.

The sum of the first (n+1) terms is $S_{n+1} = (a + ar + ar^2 + \cdots + ar^n)$.

$(1 - r)(S_{n+1}) = (a + ar + ar^2 + \cdots + ar^n - ar - ar^2 - \cdots - ar^{n+1})$, so
$(1 - r)(S_{n+1}) = a - ar^{n+1} = a(1 - r^{n+1})$.

Then $S_{n+1} = \frac{a(1-r^{n+1})}{(1-r)}$.

Since $|r| < 1$, the $\lim\limits_{n \to \infty} r^{n+1} = 0$. Therefore, we can see that as $n \to \infty$, the sum of the geometric series is $S = \frac{a}{(1-r)}$. Note the series starts with n = 0. Similar results occur if the series starts with n = 1, but of course the sum would not be the same.

For example, $\sum\limits_{n=0}^{\infty} (10)(\tfrac{1}{2})^n = \frac{10}{(1-\frac{1}{2})} = 20$.

$$\sum\limits_{n=0}^{\infty} (\tfrac{3}{4})^n = \frac{1}{(1-\frac{3}{4})} = 4.$$

$$\sum\limits_{n=0}^{\infty} (-\tfrac{1}{3})^n = \frac{1}{(1+\frac{1}{3})} = \frac{3}{4}.$$

(16) **POWER SERIES**

A power series is an infinite series of the form $\sum\limits_{n=0}^{\infty} (a_n)(x - c)^n$. This is a power series said to be centered about c. For different values of x, we have a different series. The set of x-values for which this series converges is called its interval of convergence, and will be of the form (c - R, c + R), [c - R, c + R), (c - R, c + R], or [c - R, c + R], where R is greater than or equal to zero and is called the radius of convergence. (If R = 0, the series may or may not converge at the single point c, this would have to be investigated, usually R > 0). Sometimes the series will converge for one or the other of the end-points, but this must be investigated by plugging (c - R) and (c + R) into the power series for x, and the two series investigated separately. On the interval of convergence, the power series defines a function f(x), and the series is absolutely convergent.

If we have a function f(x) which is infinitely differentiable, then we can represent f(x) on some interval by appropriate choice of the a_n. It can be shown that if $a_n = \left(\frac{f^{(n)}(c)}{n!} \right)$, the power series $f(x) = \sum\limits_{n=0}^{\infty} \left(\frac{f^{(n)}(c)}{n!} \right)(x - c)^n$ represents f(x) on some interval about x = c, which can be determined by the ratio test. This series is called a Taylor series. The notation $f^{(n)}(c)$ represents the nth derivative of f(x) evaluated at x = c, where we define $f^{(0)}(x) = f(x)$. The so-called ratio test, that we use to find the interval of convergence, says that for all x-values such that the

$$\lim_{n \to \infty} \left| \frac{a_{n+1}(x-c)^{n+1}}{a_n(x-c)^n} \right| < 1, \text{ is where the series will converge absolutely.}$$

The endpoints of the resulting x-interval need to be checked independently. Note that where this limit is less than 1 determines where the limiting ratio of the $(n+1)^{th}$ term to the $(n)^{th}$ term is like the ratio "r" of the $(n+1)^{th}$ term to the $(n)^{th}$ term of a geometric series.

<u>Example 1</u>: Find where $\sum\limits_{n=0}^{\infty} \frac{1}{(n+1)^2}(x - 3)^n$ converges.

The $\lim\limits_{n \to \infty} \left| \frac{(n+1)^2}{(n+2)^2} \frac{(x-3)^{n+1}}{(x-3)^n} \right| = |x - 3| \cdot \lim\limits_{n \to \infty} \left| \frac{(n+1)^2}{(n+2)^2} \right| < 1 \Rightarrow |x - 3| \cdot 1 < 1$

Therefore, the interval of convergence is tentatively (2, 4), we need to check the endpoints.

Plugging x = 2 into the series yields $\sum\limits_{n=0}^{\infty} \frac{1}{(n+1)^2}(-1)^n$ which is a convergent alternating series. So we will include x = 2.

Plugging x = 4 into the series yields $\sum\limits_{n=0}^{\infty} \frac{1}{(n+1)^2}$ which can be shown to be a convergent series. So we will include x = 4. Therefore, the interval of convergence is [2, 4].

Example 2: Find where $\sum\limits_{n=0}^{\infty} \frac{1}{n!} (x - 10)^n$ converges.

The $\lim\limits_{n\to\infty} \left| \frac{n!}{(n+1)!} \cdot \frac{(x-10)^{n+1}}{(x-10)^n} \right| = |x - 10| \cdot \lim\limits_{n\to\infty} \left| \frac{1}{(n+1)} \right| < 1$

which implies $|x - 10| \cdot (A) < 1$,   which implies

$|x - 10| < \frac{1}{(A)} = \infty$, because A is a quantity approaching 0.

Therefore the interval of convergence is $(-\infty, \infty) = \mathbb{R}$.

So this power series defines a function f(x) for all real numbers x.

Example 3: Find the Taylor series expansion for $f(x) = e^x$ about c = 0, and find the interval of convergence.

The function f(x) is infinitely differentiable. $f(x) = e^x$, and all of its derivatives are also e^x. Evaluating all of these functions at c = 0

yields $f^{(n)}(0) = 1$, for all n. So $\frac{f^{(n)}(0)}{n!} = \frac{1}{n!}$, for (n = 0,1,2, . . .).

Therefore $f(x) = e^x = \sum\limits_{n=0}^{\infty} \frac{1}{n!} \cdot (x - 0)^n = \sum\limits_{n=0}^{\infty} \frac{x^n}{n!}$ is a Taylor series for f(x)

$= e^x$ about c = 0 . The interval of convergence can be determined from the ratio test with results exactly like example 2, which would show that this Taylor series for $f(x) = e^x$ converges absolutely for all real numbers x. This series representation for e^x is very important and useful. We can use this series representation for e^x to approximate to any desired degree of accuracy the value of the transcendental number e (the base of the natural logarithms), by setting x = 1 in the Taylor series above.

So we have a series definition for e: $e = \sum\limits_{n=0}^{\infty} \frac{1}{n!}$.

(17) **LIMITS AND CONTINUITY**

If we have a function $f(x)$ defined on an interval $[a,b]$ or (a,b), then we can consider limits of $f(x)$ as x approaches a specific value $x = c$, where $c \in [a,b]$ or $c \in (a,b)$. The limit can be considered as x approaches c from the left, or from the right, or from both sides (unless it is one of the endpoints of a closed interval). The limits from the left and from the right are called one-sided limits, and are written $\lim_{x \to c(-)} f(x)$ and $\lim_{x \to c(+)} f(x)$ respectively. The limit from both sides is called the two-sided limit, and the two-sided limit is written $\lim_{x \to c} f(x)$. The two-sided limit exists if and only if the two one-sided limits exist and are the same.

<u>Example 1</u>: Suppose we have a function $f(x)$ defined on the interval $[1,4]$, and we consider the x-value $x = 2$.

If we're interested in the $\lim_{x \to 2(-)} f(x)$, we think of a sequence of x-values like $\{1.9, 1.99, 1.999, \ldots\}$ and we're looking for the limiting value of the sequence $\{f(1.9), f(1.99), f(1.999), \ldots\}$, if it exists.

If we're interested in the $\lim_{x \to 2(+)} f(x)$, we think of a sequence of x-values like $\{2.1, 2.01, 2.001, \ldots\}$ and we're looking for the limiting value of the sequence $\{f(2.1), f(2.01), f(2.001), \ldots\}$, if it exists.

If we're interested in the two-sided limit $\lim_{x \to 2} f(x)$, we think of an arbitrary sequence of x-values that approaches 2 from both sides simultaneously like $\{1.9, 2.1, 1.99, 2.01, 1.999, 2.001, \ldots\}$ and we're looking for the limit of the corresponding sequence of function values $\{f(1.9), f(2.1), f(1.99), f(2.01), f(1.999), f(2.001), \ldots\}$ if it exists. Again, the two-sided limit will exist if and only if the two one-sided limits exist and are equal.

At the endpoints x = 1 and x = 4,
we say that f(x) is right-continuous at x = 1 if the $\lim_{x \to 1(+)} f(x) = f(1)$.

we say that f(x) is left-continuous at x = 4 if the $\lim_{x \to 4(-)} f(x) = f(4)$.

So we have a definition for continuity on a closed interval [a,b].
A function f(x) is a continuous function on [a,b] if:

(1) f(c) is defined for every c ∈ [a,b].

(2) The $\lim_{x \to c} f(x) = f(c)$, for every c ∈ (a,b).

(3) The $\lim_{x \to a(+)} f(x) = f(a)$.

(4) The $\lim_{x \to b(-)} f(x) = f(b)$.

This leads naturally to a definition for continuity on an open interval (a,b). The endpoints a and b need not be considered. A function f(x) is continuous on (a,b) if:

(1) f(c) is defined for every c ∈ (a,b).

(2) The $\lim_{x \to c} f(x) = f(c)$, for every c ∈ (a,b).

Example 2: If f(x) = 2x² - 3 on the real numbers, and we are looking for the $\lim_{x \to 4(-)} f(x)$, the $\lim_{x \to 4(+)} f(x)$, and $\lim_{x \to 4} f(x)$, then since this is a simple polynomial function, the function values approach 2(4)² - 3 = 29 from the left and from the right of x = 4. So the two-sided limit exists and will also be 29. We would write this as the $\lim_{x \to 4} f(x) = \lim_{x \to 4} (2x^2 - 3) = 29$.

This function is a parabola and is graphed below in Figure 1. It is clear that it is continuous for all real numbers. Polynomial functions like lines, parabolas, cubics, and any nth degree polynomial function are always smooth-flowing continuous functions for all real numbers.

Figure 1

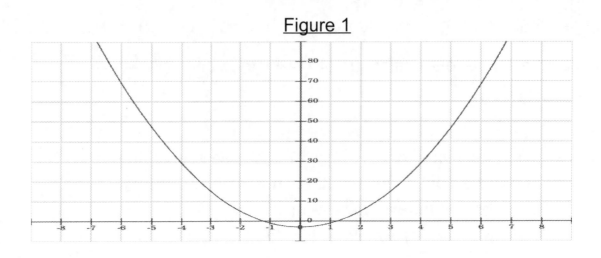

Example 3: Let $f(x) = (\frac{2}{x})$ on its domain in the real numbers.

In considering the $\lim\limits_{x \to 0(-)} (\frac{2}{x})$, we think of a sequence of x-values like

{-0.1, -0.01, -0.001, . . .} and look at the sequence of function values {f(-0.1), f(-0.01), f(-0.001), . . .} = {-20, -200, -2000, . . .}. This sequence is tending toward $-\infty$, and we write $\lim\limits_{x \to 0(-)} (\frac{2}{x}) = -\infty$.

In considering the $\lim\limits_{x \to 0(+)} (\frac{2}{x})$, we think of a sequence of x-values like

{0.1, 0.01, 0.001, . . .} and look at the sequence of function values {f(0.1), f(0.01), f(0.001), . . .} = {20, 200, 2000, . . .}. This sequence is tending toward $+\infty$, and we write $\lim\limits_{x \to 0(+)} (\frac{2}{x}) = +\infty$.

Since these two one-sided limits are different, the two-sided limit $\lim\limits_{x \to 0} (\frac{2}{x})$ doesn't exist. In fact the function f(x) is not defined at x = 0, because we would be dividing by zero. For all other real numbers c, the one-sided limits would be the same and $\lim\limits_{x \to c} (\frac{2}{x})$ would exist at x = c.

The domain of f(x) is {x | x $\in \mathbb{R}$, x $\neq 0$}. This function is graphed below in Figure 2:

42

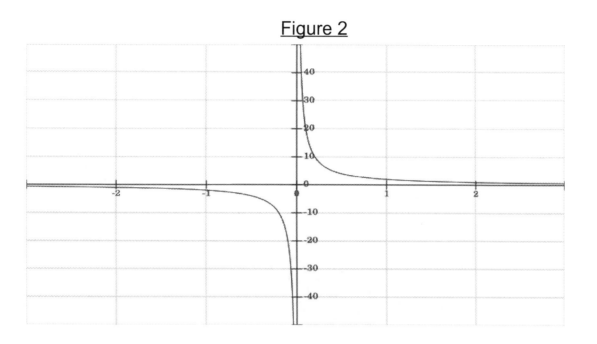

Figure 2

For f(x) = ($\frac{2}{x}$), the line x = 0 (the y-axis) is called a vertical asymptote and the line y = 0 (the x-axis) is called a horizontal asymptote. This function is a ratio of two polynomials and we call it a rational function. Vertical, horizontal (and even so-called oblique, which we won't worry about here) asymptotes occur frequently with rational functions. Vertical asymptotes usually occur at x-values where the function is not defined. Horizontal asymptotes exist when one or the other of the limits $\lim_{x \to -\infty}$ f(x) or $\lim_{x \to \infty}$ f(x) is a finite number. Polynomial functions are easy to deal with in this regard because they never have vertical or horizontal asymptotes. The end behavior of polynomials (what the function is doing as x →± ∞) is always either -∞ or ∞, depending on the degree of the polynomial and the sign of the coefficient on the highest order term.

Example 4: Another example of a rational function is f(x) = $\frac{x^2-1}{3x^2-27}$. The function will not be defined where the denominator is zero. That occurs at x = -3 and x = 3. So there will be vertical asymptotes at those

43

x-values. The $\lim\limits_{x \to \pm\infty} f(x) = \frac{1}{3}$, so $y = \frac{1}{3}$ is a horizontal asymptote. The function is continuous for all real numbers except where it is not defined at $x = \pm 3$, so its domain is $\{x \mid x \in \mathbb{R}, x \neq \pm 3\}$. This function is graphed in Figure 3:

Figure 3

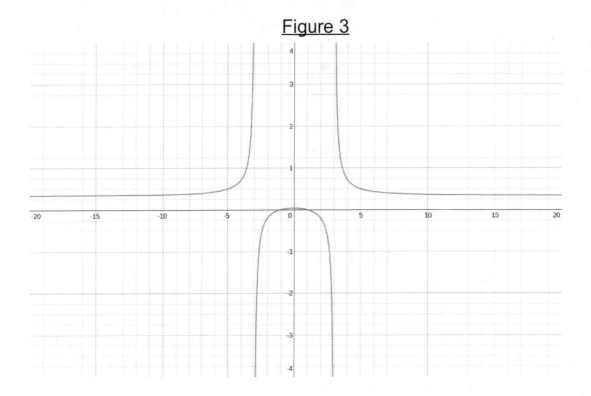

(18) **INTERMEDIATE VALUES**

Let f(x) be a function which is continuous on the interval [a,b]. Then since it is obvious that the image f([a,b]) of a continuous function y = f(x) on a closed interval [a,b] would be continuous and bounded and f([a,b]) ⊃ [f(a), f(b)] (where we can assume that f(a) < f(b)), then it would follow that the function f(x) takes all values in [f(a),f(b)]. This is known as the Intermediate Value Theorem.

(19) **THE DERIVATIVE**

The derivative of a function is one of the most important and useful tools in applied mathematics. The derivative is another function f'(x) that gives us the slope of the tangent line to a function f(x) at the point (x,f(x)) on a certain interval (a,b). The functions f(x) that we will consider in this section will be assumed to be continuous and differentiable on certain intervals, differentiable meaning that the tangent line on the curve y = f(x) exists and is unique at all points on some interval. The tangent line would not be unique where f(x) has a cusp (a sharp point). f(x) is said to be smooth where there are no cusps. A good example of a cusp that you should be familiar with happens with the function f(x) = $|x|$. This function has a cusp at the origin and there is no unique tangent line there. The slope of the tangent line at a point on the curve y = f(x) tells us the instantaneous rate of change of f(x) with respect to x, at that point.

If we have a function y = f(x) continuous on (a,b), the derivative of f(x), denoted in different ways in different contexts ($\frac{dy}{dx}$, f'(x), y', $D_x f(x)$) is defined as the $\lim_{\Delta x \to 0}$ ($\frac{\Delta y}{\Delta x}$) if it exists. The increment of the dependent variable y, written Δy, is Δy = f(x + Δx) - f(x). Δx is simply an increment of the independent variable x. Note that the derivative is a limit like any other (but an important one) and we need the left-hand and right-hand limits to be the same. Hence you can see why it is not defined at a cusp.

When the slope of a tangent line to a function f(x) is positive at a certain point, then f(x) is increasing as x increases (as we go from left to right) at that point. If f'(a) and f'(b) are positive and $f'(b) > f'(a)$, then the function is increasing faster at the point (b,f(b)) than it is at (a,f(a)).

When the slope of a tangent line to a function f(x) is negative at a certain point, then f(x) is decreasing as x increases (as we go from left to right) at that point. If f'(a) and f'(b) are negative and $|f'(b)| > |f'(a)|$, then the function is decreasing faster at the point (b,f(b)) than it is at (a,f(a)).

Some functions, like a line with a positive slope, is increasing everywhere. If we have a line with a negative slope, it is decreasing everywhere. For a parabola that opens upward, it is decreasing everywhere to the left of its vertex and its derivative would be negative at points there, and it is increasing everywhere to the right of its vertex and its derivative would be positive at points there. This is what the derivative tells us, how the function is changing and at what rate, at a specific point in its domain. It is this information that we exploit in order to analyze functions.

<u>Example 1</u>: If y = f(x) = 4x² + x + 3 on (-∞, ∞), find the derivative f'(x).

$$f'(x) = \lim_{\Delta x \to 0} \left(\frac{\Delta y}{\Delta x} \right) = \lim_{\Delta x \to 0} \frac{f(x + \Delta x) - f(x)}{\Delta x}$$

$$= \lim_{\Delta x \to 0} \frac{4(x + \Delta x)^2 + (x + \Delta x) + 3 - (4x^2 + x + 3)}{\Delta x}$$

$$= \lim_{\Delta x \to 0} \frac{4x^2 + 8x(\Delta x) + 4(\Delta x)^2 + x + (\Delta x) + 3 - 4x^2 - x - 3}{(\Delta x)}$$

$$= \lim_{\Delta x \to 0} \frac{8x(\Delta x) + 4(\Delta x)^2 + (\Delta x)}{(\Delta x)}$$

$$= \lim_{\Delta x \to 0} (8x + 1 + 4(\Delta x))$$

As $\Delta x \to 0$, the last term approaches 0. So in the limit f'(x) = 8x + 1.

The instantaneous rate of change of f(x) = 4x² + x + 3, when x = -2, is f'(-2) = 8(-2) + 1 = -15

The instantaneous rate of change of f(x) = 4x² + x + 3, when x = 3, is f'(3) = 8(3) + 1 = 25.

These instantaneous rates of change are the slope of tangent lines at the points (-2,17) and (3,42), shown along with f(x) in Figure 4:

Figure 4

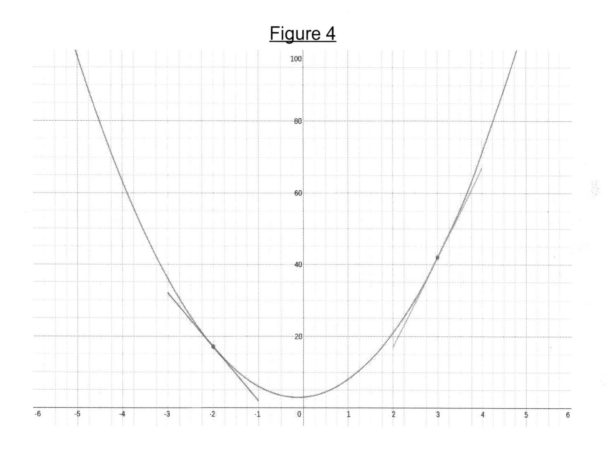

Example 2: If y = f(x) = $\sqrt{x}$, on (0, ∞). Find f'(x).

$$f'(x) = \lim_{\Delta x \to 0} \left(\frac{\Delta y}{\Delta x}\right) = \lim_{\Delta x \to 0} \frac{f(x + \Delta x) - f(x)}{\Delta x}$$

$$= \lim_{\Delta x \to 0} \frac{\sqrt{x + \Delta x} - \sqrt{x}}{\Delta x}$$

$$= \lim_{\Delta x \to 0} \frac{\sqrt{x + \Delta x} - \sqrt{x}}{\Delta x} \cdot \frac{\sqrt{x + \Delta x} + \sqrt{x}}{\sqrt{x + \Delta x} + \sqrt{x}}$$

$$= \lim_{\Delta x \to 0} \frac{x + \Delta x - x}{(\Delta x)(\sqrt{x + \Delta x} + \sqrt{x})}$$

$$= \lim_{\Delta x \to 0} \frac{1}{(\sqrt{x + \Delta x} + \sqrt{x})}$$

As $\Delta x \to 0$, $\sqrt{x + \Delta x}$ simply approaches $\sqrt{x}$.
So in the limit, $f'(x) = \frac{1}{2\sqrt{x}}$.

The instantaneous rate of change, for $f(x) = \sqrt{x}$, when x = 4, is
$f'(4) = (1/(2\sqrt{4})) = (\frac{1}{4})$.
The instantaneous rate of change, for $f(x) = \sqrt{x}$, when x = 16, is
$f'(16) = (1/(2\sqrt{16})) = (\frac{1}{8})$.
These instantaneous rates of change are the slope of tangent lines at
the points (4,2) and (16,4), and are shown along with f(x), in Figure 5.
Note that $f(x) = \sqrt{x}$ is increasing over its domain $[0, \infty)$, but the rate of
increase slows as x increases:

Figure 5

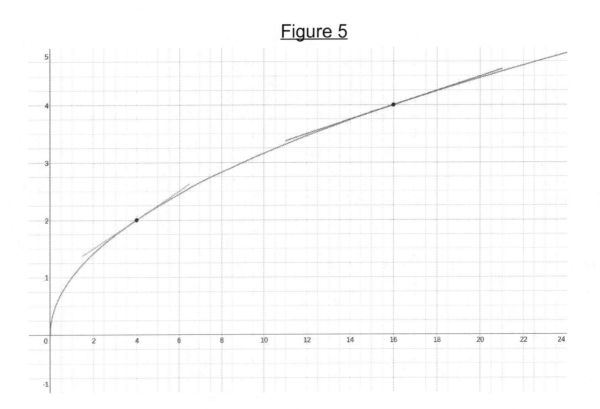

Example 3: If y = f(x) = ($\frac{1}{x}$), on (-∞ , 0) ∪ (0, ∞), find f'(x).

$$f'(x) = \lim_{\Delta x \to 0} (\frac{\Delta y}{\Delta x}) = \lim_{\Delta x \to 0} (\frac{1}{\Delta x})(\frac{1}{x + \Delta x} - \frac{1}{x})$$

$$= \lim_{\Delta x \to 0} (\frac{1}{\Delta x})(\frac{x - (x + \Delta x)}{(x + \Delta x)(x)})$$

$$= \lim_{\Delta x \to 0} (\frac{-1}{(x + \Delta x)(x)})$$

So in the limit, f'(x) = ($\frac{-1}{x^2}$)

The instantaneous rate of change, for f(x) = $\frac{1}{x}$, when x = - $\frac{1}{2}$, is
f'(- $\frac{1}{2}$) = ($\frac{-1}{1/4}$) = -4.
The instantaneous rate of change, for f(x) = $\frac{1}{x}$, when x = $\frac{1}{2}$, is
f'($\frac{1}{2}$) = ($\frac{-1}{1/4}$) = -4.

These instantaneous rates of change are the slope of tangent lines at
the points (- $\frac{1}{2}$,- 2) and ($\frac{1}{2}$, 2), shown along with f(x), in Figure 6. Note
that this function is decreasing everywhere that it is defined.

Figure 6

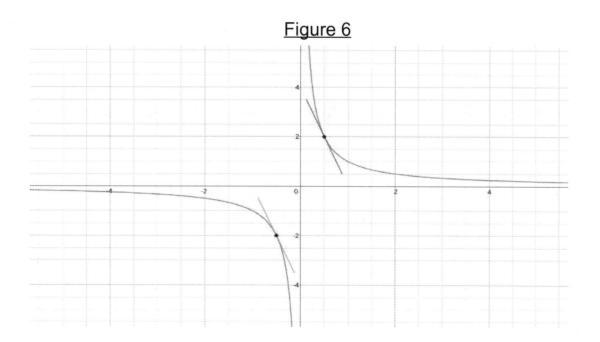

There are techniques for calculating derivatives that you learn and get practice with in the first year of calculus, which it is assumed the reader has taken. For review purposes we will list some rules for finding derivatives along with the derivative of some common functions. The notation $\frac{d}{dx}$ means the derivative with respect to x, which is what we have found in the three examples above.

For a constant c, the $\frac{d}{dx}$ (c) = 0.

The $\frac{d}{dx}$ (a · f(x)) = a · $\frac{d}{dx}$ f(x).
The $\frac{d}{dx}$ (a · f(x) + b · g(x)) = a · $\frac{d}{dx}$ f(x) + b · $\frac{d}{dx}$ g(x).

Because of these last two rules, we call differentiation a linear operator.

<u>The Power Rule</u>: $\frac{d}{dx}$ (ax^n) = nax^{n-1} , a and n are any real numbers.

<u>The Product Rule</u>: If f(x) and g(x) are differentiable functions and h(x) = f(x)g(x), then h'(x) = f(x)g'(x) + g(x)f'(x).

<u>The Quotient Rule</u>: If f(x) and g(x) are differentiable functions and h(x) = f(x)/g(x), then h'(x) = $(\frac{1}{(g(x))^2})$ (g(x)f'(x) - f(x)g'(x))

<u>The Chain rule</u>: For a composite function y = f(g(x)), let u = g(x),
then $\frac{dy}{dx} = \frac{dy}{du} \cdot \frac{du}{dx}$.

<u>Example 1</u>: If y = f(x) = 7(x^2 + 3x)^3, we let u = x^2 + 3x.
Then y = 7u^3, $\frac{dy}{du}$ = 21u^2, and $\frac{du}{dx}$ = 2x + 3.
Therefore $\frac{dy}{dx}$ = (21u^2) · (2x + 3) = 21(x^2 + 3x)^2(2x + 3).

<u>Example 2</u>: If y = f(x) = e$^{(sinx)}$, we let u = sinx.
Then y = e^u, $\frac{dy}{du}$ = e^u, and $\frac{du}{dx}$ =cosx.

Therefore $\frac{dy}{dx} = (e^u)(\cos x) = e^{\sin x}(\cos x)$.

Implicit differentiation is another technique useful for deriving an expression for the derivative. In certain situations where we have an equation in x and y, it may be impossible (or at least very difficult) to rearrange the equation to get y as an explicit function of x from which we could find the derivative in a straightforward way. What we do is assume that y is an implicit function of x, and write $\frac{dy}{dx}$ whenever we're differentiating the variable y with respect to x. Therefore this technique uses the chain rule (because we multiply by $\frac{dy}{dx}$ when we assume y is some unknown implicit function of x) along with the other usual rules of differentiation. Two examples:

Example 1: Suppose $x^2y^3 + \sin(xy) + 4xy = (\pi^3 + 16\pi + 8)/8$. Find $\frac{dy}{dx}$.
We simply, from left to right, differentiate the equation with respect to x, writing $\frac{dy}{dx}$ everywhere we differentiate y with respect to x. Then we solve the resulting equation for $\frac{dy}{dx}$. Notice how the product rule and the chain rule are used because we are assuming y is a function of x.

$(x^2)(3y^2)\frac{dy}{dx} + (y^3)(2x) + \cos(xy)(x\frac{dy}{dx} + y(1)) + 4(x\frac{dy}{dx} + y(1)) = 0$
(collecting terms with $\frac{dy}{dx}$ on the left, and the others on the right)
$((x^2)(3y^2) + \cos(xy)(x) + 4x) \cdot \frac{dy}{dx} = -2xy^3 - y\cos(xy) - 4y$

So: $\frac{dy}{dx} = \frac{(-2xy^3 - y\cos(xy) - 4y)}{(3x^2y^2 + x\cos(xy) + 4x)}$.

Notice that the final expression for $\frac{dy}{dx}$ involves both x and y. If we know that the curve represented by the given equation goes through the point $(1, \pi/2)$, then plugging x = 1 and y = $\frac{\pi}{2}$, we find that at this

51

point $\frac{dy}{dx} \approx (-1.23)$.

Example 2: Suppose $(x^3 + 3x^2)y^2 - x\sqrt{y} + ye^x = 1$. Find $\frac{dy}{dx}$.

$(x^3 + 3x^2)2y\frac{dy}{dx} + y^2(3x^2 + 6x) - (x \cdot (\frac{1}{2\sqrt{y}})\frac{dy}{dx} + \sqrt{y}) + (ye^x + e^x(\frac{dy}{dx})) = 0$

(collecting terms with $\frac{dy}{dx}$ on the left, and the others on the right)

$\Rightarrow (2y(x^3 + 3x^2) - (\frac{x}{2\sqrt{y}}) + e^x) \cdot \frac{dy}{dx} = (-y^2(3x^2 + 6x) + \sqrt{y} - ye^x)$

So: $\frac{dy}{dx} = \dfrac{(-y^2(3x^2 + 6x) + \sqrt{y} - ye^x)}{(2y(x^3 + 3x^2) - (\frac{x}{2\sqrt{y}}) + e^x)}$

If we know that the curve represented by the given equation goes through the point (0,1), then plugging x = 0 and y = 1, we find that at this point $\frac{dy}{dx}$ = 0.

The derivative of some common functions:

$\frac{d}{dx}(sinx) = cosx.$ $\qquad \frac{d}{dx}(cosx) = -sinx.$ $\qquad \frac{d}{dx}(tanx) = sec^2x.$

$\frac{d}{dx}(cscx) = -cscxcotx.$ $\qquad \frac{d}{dx}(secx) = secxtanx.$ $\qquad \frac{d}{dx}(cotx) = -csc^2x.$

$\frac{d}{dx}(lnx) = (\frac{1}{x}).$ $\qquad\qquad\qquad \frac{d}{dx}(e^x) = e^x.$

Examples of the derivatives of some functions:

(1) $y = 3x^4 - 2x^9 - \sqrt{x} \Rightarrow \frac{dy}{dx} = 12x^3 - 18x^8 - (\frac{1}{2\sqrt{x}})$ $\qquad$ (power rule)

(2) $y = 2x + cos(20x) \Rightarrow \frac{dy}{dx} = 2 - 20sin(20x)$ $\qquad$ (chain rule)

(3) $y = ln(2x^2 + 35x) \Rightarrow \frac{dy}{dx} = (\frac{1}{(2x^2 + 35x)}) \cdot (4x + 35)$ $\quad$ (chain rule)

(4) $y = secx - e^{-x} \Rightarrow \frac{dy}{dx} = secxtanx + e^{-x}$ $\qquad$ (chain rule)

(5) $y = 2x \cdot tanx \Rightarrow \frac{dy}{dx} = (2x)(sec^2x) + (tanx)(2)$ (product rule)

(6) $y = \frac{3x}{(4x + 2)} \Rightarrow \frac{dy}{dx} = \frac{(4x + 2)(3) - (3x)(4)}{(4x + 2)^2} = \frac{6}{(4x + 2)^2}$ (quotient rule)

(7) $y = xe^{-3x} \Rightarrow \frac{dy}{dx} = (x)(-3e^{-3x}) + (e^{-3x})$ $\qquad$ (product rule)

(20) **MORE ABOUT THE DERIVATIVE**

Throughout the last few centuries there have been many uses of the derivative in mathematics and the sciences, and one couldn't possibly attempt to mention them all. However, in this section we will cover four topics involving derivatives that are very important in solving problems in mathematics and its applications: related rates, linear approximation, optimization problems, and physical motion.

Related Rates problems are a variation on the implicit differentiation idea. We now assume that every variable is an implicit function of time. So we derive an expression, for all the variables involved, relating all the different rates of change with respect to time in a single equation.

Example 1: Suppose we have a right triangle with legs of length x and y, and with hypotenuse z. We know that $x^2 + y^2 = z^2$. The rates of change of x and y with respect to time t are $\frac{dx}{dt}$ = -3 ft/sec (x is decreasing) and $\frac{dy}{dt}$ = 5 ft/sec (y is increasing) at the instant that x = 55 ft and y = 110 ft. What is $\frac{dz}{dt}$? At that instant, z will be $\sqrt{x^2 + y^2}$ from the pythagorean theorem. This means that $z = \sqrt{(55)^2 + (110)^2}$, or $z = \sqrt{15125} \approx 123$ ft. From the relation $x^2 + y^2 = z^2$, when we differentiate from left to right all variables with respect to time t (this will involve the chain rule since we are assuming that x, y, and z are implicit functions of time), we get

$$2x\frac{dx}{dt} + 2y\frac{dy}{dt} = 2z\frac{dz}{dt} \ .$$

Now we just substitute x = 55, y = 110, z = 123, $\frac{dx}{dt}$ = -3, $\frac{dy}{dt}$ = 5, into

$$\frac{dz}{dt} = \frac{2x \frac{dx}{dt} + 2y \frac{dy}{dt}}{2z} \text{ , resulting in}$$

$$\frac{dz}{dt} = \frac{2(55)(-3) + 2(110)(5))}{2(123)} \approx 3.13 \text{ ft/sec}$$

.

Example 2: We have an expandable right circular cylinder that at a specific point in time is of height h = 63 mm, and base radius r = 21mm. At that specific point in time, $\frac{dr}{dt}$ = 2 mm/sec and $\frac{dh}{dt}$ = 3.5 mm/sec, and we are interested in what the rate of change of the volume V is with respect to time. From V = ($\pi r^2 h$), we find using the chain rule and the product rule, since all variables here are implicit functions of time,

$$\frac{dV}{dt} = \pi (r^2 (\tfrac{dh}{dt}) + h(2r(\tfrac{dr}{dt})))$$

So $\frac{dV}{dt} = \pi ((21)^2(3.5) + (63)(2)(21)(2)) \approx 21,474.36 \text{ mm}^3/\text{sec}$.

Increments and Differentials: Suppose we have a function y = f(x), and its derivative $\frac{dy}{dx}$ = f'(x). Multiplying through by dx, we get dy = f'(x)dx.

At a point (c,f(c)) on the curve, f'(c) is the slope of the tangent line. For the independent variable x, $\Delta x = dx$. For the dependent variable y, dy = f'(c)dx. The actual change in y from (c) to (c + dx) is Δy , and it is approximated by dy, that is $\Delta y \approx dy$. This idea leads us to the use of a linear approximation.

Linear Approximation is a very common and useful method of approximating a function value. We take advantage of the fact that the derivative gives us the slope of a function at a point (a,f(a)) on a curve. With knowledge of the slope at the point (a,f(a)), we can find the equation of the tangent line at (a,f(a)). The function values along this tangent line are then a good approximation to the function values for f(x) in a small neighborhood around x = a. Let me illustrate how this can be done with two examples.

Example 1: Suppose we have the function f(x) = $\sqrt{x}$, and we wish to estimate the $\sqrt{4.001}$. The derivative of the function f(x) is f'(x) = ($\frac{1}{2\sqrt{x}}$). We can estimate $\sqrt{4.001}$ very accurately because we know the square root of 4 exactly, and 4.001 is very close to 4.

From the equation of a line y = mx + b at x = a, let y = L(x), b = f(a), m = f'(a), and let x be Δx = (x - a). So we have the equation for the tangent line at (a, f(a)):   L(x) = f(a) + f'(a)(x - a). Use this equation for x = 4.001 and a = 4 to get:

$$\sqrt{4.001} \approx L(4.001) = \sqrt{4} + (\frac{1}{2\sqrt{4}})(4.001 - 4)$$
$$\sqrt{4.001} \approx 2 + (\frac{1}{4})(0.001) = 2.00025.$$

The value I get from my scientific calculator is 2.000249984.

Example 2: The volume of a sphere is ($\frac{4}{3}$) πr^3. Estimate the volume of a sphere if r = 3.001. V(r) = ($\frac{4}{3}$) πr^3 and $\frac{dV}{dr}$ = 4 πr^2.

$$V(3.001) \approx L(3.001) = V(3) + V'(3)(3.001 - 3)$$
$$V(3.001) \approx (\frac{4}{3}) \pi (3)^3 + (4 \pi (3)^2)(0.001)$$
$$V(3.001) \approx 113.2104328.$$

The value I get from my scientific calculator is 113.2104706.
The linear approximation is accurate to four decimal places.

Optimization problems that we will consider here are those where we have a function y = f(x) with a curve associated with it, and where we are interested in finding relative maximums and minimums, and absolute maximums and minimums on a closed interval [a,b]. We will assume that the function is continuous on [a,b] and differentiable on (a,b).

To find relative maximums and minimums we usually consider where $f'(x) = 0$ in the interior of [a,b] (called critical values), because this is where we have a horizontal tangent line, indicating that at that point we are at the top of a peak of the curve (a rel-max), or at the bottom of a valley of the curve (a rel-min), or at a saddle point (neither a rel-max or a rel-min).

To find absolute maximums and minimums we are interested in where $f'(x) = 0$ in the interior of [a,b] (the critical values) and also the function values at the boundary of the interval. Once we have the x-values where $f'(x) = 0$, we can use $f''(x)$ to find out if we have a rel-max or a rel-min there. If x = c is a critical value, there will be a rel-max at c if $f''(c) < 0$, there will be a rel-min at c if $f''(c) > 0$, and there will be neither (called a saddle point) if $f''(c) = 0$.

$f''(x)$ is the second derivative, the derivative of $f'(x)$, and tells us how $f'(x)$ is changing in exactly the same way that $f'(x)$ tells us how $f(x)$ is changing. When we have a rel-max at x = c, ($f''(c) < 0$), $f'(x)$ is positive and then zero and then negative as it goes over the rel-max, that is, $f'(x)$ is decreasing. When we have a rel-min at x = c, ($f''(c) > 0$), $f'(x)$ is negative and then zero and then positive as it goes through the rel-min, that is, $f'(x)$ is increasing.

Example 1: Find the rel-max's, rel-min's, abs-max's, and abs-min's for $y = f(x) = x^3 - 5x^2 + 2x - 1$ on [-5,5].

$f'(x) = 3x^2 - 10x + 2$
$f''(x) = 6x - 10$

Using the quadratic formula, we find that $f'(x) = 0$ at x = 0.207 and at x = 3.113. These are the critical values.

$f''(0.207) < 0$ and $f''(3.113) > 0$

f(0.207) = (-0.7914) and f(3.113) = (-13.06)
Using this information, we can see that f(x) has a rel-max at (0.207,-0.7914), and f(x) has a rel-min at (3.113,-13.06).

At the boundaries, f(-5) = -261, and f(5) = 9. So f(x) has an abs-min of (-261) at x = -5, and an abs-max of (9) at x = 5. So on the closed interval [-5 5], the function has its abs-min and abs-max at neither of the critical values in the interior, but rather on the boundaries. The graph of this situation is in Figure 7:

Figure 7

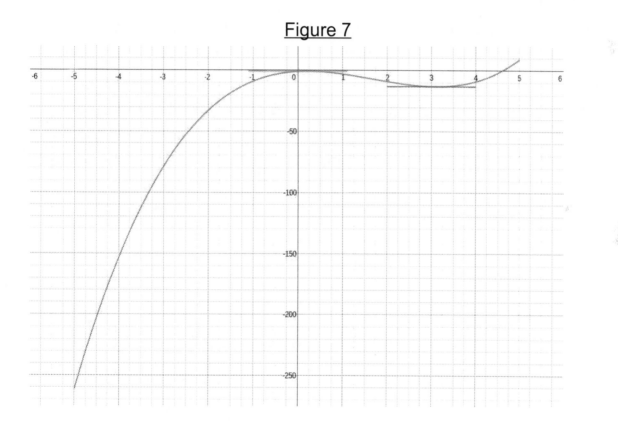

Example 2: Suppose we have a box with a square base (length of side x) and a height y, that must have a volume of 36 ft³. We will say that x is in [1,6]. Show that the box of minimum surface area occurs if the box is

a cube.

The volume $V = x^2y = 36$. The surface area $S(x,y) = 2x^2 + 4xy$. Since $V = x^2y = 36$, then $y = \frac{36}{x^2}$. Plugging this into the formula for S gives $S(x) = 2x^2 + 4x(\frac{36}{x^2}) = 2x^2 + (\frac{144}{x})$, making S a function of one variable.

Then $S'(x) = 4x - (\frac{144}{x^2})$ and $S''(x) = 4 + (\frac{288}{x^3})$.
Setting $S'(x) = 4x - (\frac{144}{x^2}) = 0$ to find the critical values yields
$4x^3 = 144 \Rightarrow x^3 = 36 \Rightarrow x = \sqrt[3]{36}$.
So $y = \frac{36}{x^2} = \frac{36}{(\sqrt[3]{36})^2} = \sqrt[3]{36}$. So x = y, which says that the surface

area is minimized when the box is a cube (where each side is $\sqrt[3]{36} \approx 3.3$ ft). This value satisfies the constraint that the volume must be 36 ft³. The fact that $S''(\sqrt[3]{36}) > 0$ confirms that the surface area is minimized when we have a cube. The graph of S(x) is in Figure 8:

Figure 8

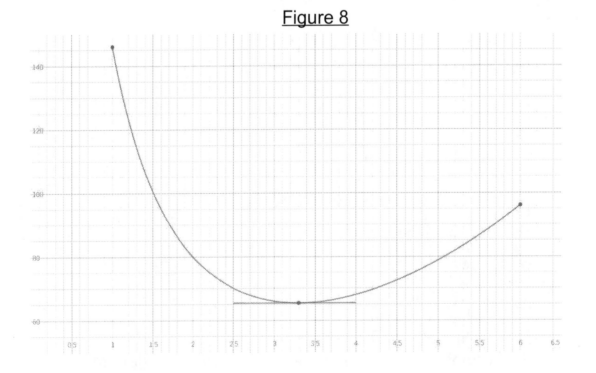

Example 3: Find two numbers x and y in [0,90] such that xy is

maximized subject to the constraint that x + y = 90.

Let P(x,y) = xy. The constraint says that y = 90 - x, so we can get P(x,y) as a function of only one variable. P(x) = x(90 - x) = -x² + 90x.

P'(x) = -2x + 90 = 0 ⇒ the critical value is x = 45. P"(x) = -2, that is, P"(x) is negative for every x in [0,90], so we have a rel-max at x = 45.

So x = 45 and y = 45 are the two numbers whose product is a maximum subject to the constraint that their sum must be 90. This situation is shown in Figure 9.

Figure 9

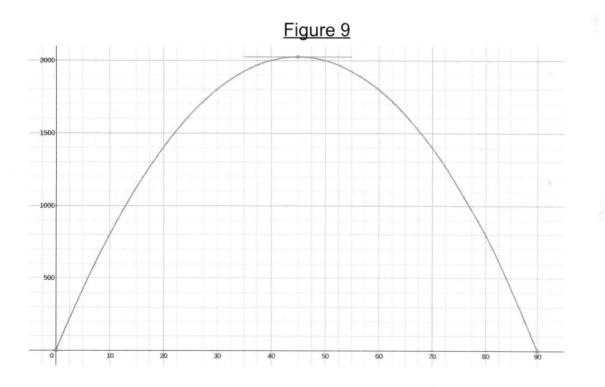

Physical Motion: For many years, physics and mathematics has developed hand in hand, so there are many applications of calculus in physics. One of the most important is physical motion involving the concepts of position, velocity, and acceleration.

When a particle is in motion with position function x(t), the average rate of change in position (or average velocity) during a time interval Δt is defined to be $\Delta v = (\frac{\Delta x}{\Delta t})$. As $\Delta t \to 0$, we get the derivative of x(t) with respect to time t, which is the velocity function v(t). So we have:

$$v(t) = \frac{dx}{dt} \, .$$

The average rate of change of velocity (or average acceleration) during a time interval Δt is defined to be $\Delta a = (\frac{\Delta v}{\Delta t})$. As $\Delta t \to 0$, we get the derivative of v(t) with respect to time t, which is the acceleration function a(t). So we have:

$$a(t) = \frac{dv}{dt} \, .$$

For example, if we have a particle which is moving in one dimension with constant acceleration a, and having a position function

$$x(t) = x_0 + v_0 t + \tfrac{1}{2} at^2$$

where x_0 and v_0 are the initial position and initial velocity respectively, then the velocity function is:

$$v(t) = \frac{dx}{dt} = v_0 + at \, .$$

It follows that the acceleration function is:

$$a(t) = \frac{dv}{dt} = a \, .$$

These are exactly the kinematics equations for an object moving vertically near the earth's surface with constant acceleration a = -9.8 meters/ sec^2 (neglecting air resistance). Usually this is found in the first chapter of any physics textbook. Calculus is involved in virtually all aspects of classical and modern physics.

(21) **ROLLE'S THEOREM**

If f(x) is a continuous function on [a,b] and differentiable on the open interval (a,b), and f(a) = f(b) = 0, then there exists a number c in (a,b) such that f'(c) = 0.

Proof: We will disregard the special case of f(x) = 0 on [a,b]. Without loss of generality, we can consider the case that there exists a number z in (a,b) such that f(z) > 0. Assume that there does not exist a number c in (a,b) such that f'(c) = 0. Then f(x) would be monotonically increasing and f(b) would be greater than zero. This is a contradiction of our hypothesis that f(b) = 0, so we must conclude that there exists a c in (a,b) such that f'(c) = 0.

(22) **THE MEAN VALUE THEOREM**

If f(x) is continuous on [a,b] and differentiable on (a,b), then there exists a number c in (a,b) such that $f'(c) = \frac{f(b)-f(a)}{b-a}$.

Proof: The line through the two points (a,f(a)) and (b,f(b)) has slope

$\frac{f(b)-f(a)}{b-a}$. So the equation of this line is g(x) = $(\frac{f(b)-f(a)}{b-a}) \cdot$ (x - a) + f(a).

Then h(x) = (f(x) - g(x)) = f(x) - $(\frac{f(b)-f(a)}{b-a}) \cdot$ (x -a) - f(a).

Note that h(a) = h(b) = 0. Therefore, from Rolle's theorem, there exists a number c in (a,b) such that h'(c) = (f'(c) - g'(c)) = f'(c) - $(\frac{f(b)-f(a)}{b-a})$ = 0. Therefore, it follows that f'(c) = $(\frac{f(b)-f(a)}{b-a})$, which we wished to show.

(23) ANTIDERIVATIVES AND INDEFINITE INTEGRALS

A function $F(x)$ such that $F'(x) = f(x)$, is called an antiderivative of $f(x)$.

We write this as $F(x) = \int f(x)dx$ and call this an indefinite integral. $f(x)$ is called the integrand.

Another way of writing $F(x)$ is $F(x) = \int_a^x f(t)dt$. We will discuss this in more detail in the next section, but $\int_a^x f(t)dt$ is called a definite integral and is the area under $f(x)$ and above the x-axis, and between 'a' and 'x', where the upper limit x is the variable, which can vary. This definite integral can be interpreted as the accumulation of area as x increases. We call (t) a dummy variable because we could have chosen any symbol instead of (t). We then say that $F'(x) = \frac{d}{dx} \int_a^x f(t)dt = f(x)$. This will be proven rigorously when we discuss the Fundamental Theorem of Calculus. What we should notice here is that the rate of change of the accumulation of area, as x increases, depends on the value of $f(x)$. If $f(x)$ is small, then area is accumulating slowly, and it is accumulating faster when $f(x)$ is large.

We note that $\int f(x)dx = F(x) + C$, where C is an arbitrary constant. This is true because for any constant C, $\frac{d}{dx}(F(x) + C) = \frac{d}{dx}F(x) + \frac{d}{dx}(C) = f(x)$, because the derivative of any constant is zero. This says that there are an infinite number of antiderivatives for $f(x)$, but they all differ at most by a constant. As a special case, $\int dx = x + C$, since $\frac{d}{dx}(x + C) =$

$\frac{d}{dx}$ (x) + $\frac{d}{dx}$ (C) = 1, which is the integrand in $\int dx = \int (1)dx$.

Just as differentiation is a linear operator, antidifferentiation is a linear operator also, meaning that:

$\int af(x)dx = a \int f(x)dx$, for any constant a, and

$\int (af(x)dx + bg(x)dx) = a\int f(x)dx + b\int g(x)dx$, for any functions f(x) and g(x), and any constants a and b.

When we have a derivative formula, then we can derive an integral formula. For example if $\frac{d(sinx)}{dx}$ = cosx, then d(sinx) = (cosx)dx. Taking the integral of both sides: $\int d(sinx) = \int cosx\ dx$. This gives us the integral formula $\int cosx\ dx = sinx + C$, where C is an arbitrary constant.

The antiderivative of some common functions:

$\int x^n dx = (\frac{x^{n+1}}{n+1}) + C$, (n ≠ -1). $\qquad \int (\frac{1}{x}) dx = ln|x| + C$.

$\int e^x dx = e^x + C$. $\qquad \int sinx\ dx = -cosx + C$.

$\int cosx\ dx = sinx + C$. $\qquad \int sec^2x\ dx = tanx + C$.

$\int csc^2x\ dx = -cotx + C$. $\qquad \int secxtanx\ dx = secx + C$.

$\int cscxcotx\ dx = -cscx + C$. $\qquad \int \frac{1}{\sqrt{1-x^2}}\ dx = sin^{-1}(x) + C$.

$\int \frac{1}{1+x^2}\ dx = tan^{-1}(x) + C$. $\qquad \int \frac{1}{x\sqrt{x^2-1}}\ dx = sec^{-1}(x) + C$.

Integration by Parts:

There are many techniques of integration that you learn in your calculus course, and I have included the indefinite integrals (antiderivatives) of some common functions. There is one technique of integration that we should include because it is so important and useful: it is called integration by parts.

From the product rule for differentiation where we have the product of two functions of x, u(x) and v(x),

$$\frac{d(u(x)v(x))}{dx} = u(x)\frac{d(v(x))}{dx} + v(x)\frac{d(u(x))}{dx} ,$$

which leads to the differential formulation after multiplying through by dx,

$$d(u(x)v(x)) = u(x)\,d(v(x)) + v(x)\,d(u(x)).$$

Integrating this equation, we get

$$\int d(u(x)v(x)) = \int u(x)d(v(x)) + \int v(x)d(u(x)).$$

Then letting u = u(x), v = v(x), we get the common representation:

$$\int udv = uv - \int vdu.$$

The aspect of integration by parts which can be difficult in some cases is the correct choice of what part of the integrand will be u and what part will be dv. Lets work two examples using this technique.

<u>Example 1</u>: Suppose we wish to solve $\int x\,(e^x)\,dx$. If we let u = x and

dv = $e^x\,dx$, then du = dx and v = $\int e^x\,dx = e^x$. We have all the pieces

(u, v, du, and dv), so then $\int x\,(e^x)\,dx = (x)(e^x) - \int e^x dx = x\,e^x - e^x + C$

$= e^x(x-1) + C$.

<u>Example 2</u>: Evaluate $\int x\ln(x)\,dx$. Let's choose u = ln(x), and dv = xdx.

Then du = $(\frac{1}{x})\,dx$, and v = $\frac{x^2}{2}$. So then we have: $\int x\ln(x)\,dx =$

$(\frac{x^2}{2})\ln(x) - \frac{1}{2}\int x dx$, which says that $\int x\ln(x)\,dx = (\frac{x^2}{2})\ln(x) - \frac{x^2}{4} + C$.

<u>Examples of antiderivatives</u>. (C is an arbitrary constant)

(1) $\int (8x^3 + 2e^x)dx = 2x^4 + 2e^x + C$

(2) $\int (8\sin x + \frac{2}{x})dx = -8\cos x + 2\cdot\ln|x| + C$

(3) $\int (8e^x + 12x^{-3})dx = 8e^x + \frac{12x^{-2}}{(-2)} + C = 8e^x - \frac{6}{x^2} + C$

(4) $\int (\csc^2 x - 9x - 10)dx = -\cot x - \frac{9x^2}{2} - 10x + C$

(5) $\int (\frac{1}{x} + \cos x - e^x + 1) = \ln|x| + \sin x - e^x + x + C$

(6) $\int (1 + x + x^2 + x^3)\,dx = x + \frac{x^2}{2} + \frac{x^3}{3} + \frac{x^4}{4} + C$

(7) $\int (\frac{1}{2\sqrt{x}})\,dx = \frac{1}{2}\int (x^{-\frac{1}{2}})\,dx = (\frac{1}{2})\frac{x^{\frac{1}{2}}}{\frac{1}{2}} = \sqrt{x} + C$

(24) **THE DEFINITE INTEGRAL**

In many applications of calculus, for a function f(x) which is continuous on a closed interval [a,b], we are interested in the area between the x-axis and the function f(x), and between the vertical lines x = a and x = b. This area usually has some physical meaning, or it may represent a probability, etc. It suffices to assume at this point that f(x) $\geq$ 0 on [a,b].

We denote this area by $\int_a^b$ f(x) dx, and we call this a definite Integral.

Figure 10 shows the $\int_{-2}^{4}$ $\sqrt{16-x^2}$ dx . This definite integral is equal to the shaded area.

Figure 10

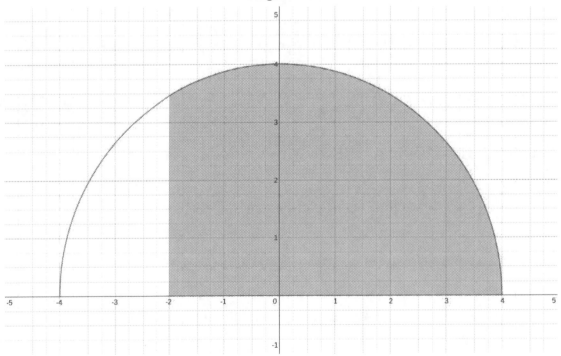

The way that this area is found is by first approximating it with a sum of n rectangles on [a,b] that each have a base of equal length $\Delta x_i = \left(\frac{b-a}{n}\right)$ and height $f(w_i)$, where w_i is any point in Δx_i, for i = 1,2,...,n. This sum is written $\sum_{i=1}^{n} f(w_i)(\Delta x_i)$, and for each n, it approximates the exact area under the curve with the sum of the areas of the n rectangles. This sum is called a Riemann sum, after the 19th century mathematician Bernhard Riemann. The exact area is found by refining the partition of the x-axis between a and b, so that the number of rectangles n gets larger and larger. The limit of the sum as $n \to \infty$, $\lim_{n\to\infty} \sum_{i=1}^{n} f(w_i)(\Delta x_i)$, will exist if f(x) is continuous on [a,b] and it equals the exact value denoted by $\int_{a}^{b} f(x)\, dx$.

Example 1: As an example of finding the area under a curve using the Riemann Sum approach, I will use it to evaluate $\int_{0}^{1} x^2\, dx$. The reader, which we are assuming has taken a first course in calculus, should be able to readily evaluate this integral using the Fundamental Theorem of Calculus (FTC) and find that it equals $\frac{1}{3}$. We will review this very important theorem in the next few sections. However it's an instructive exercise to do this using a Riemann sum.

Some summation formulas which often prove very useful in this type of problem are:

(1) $\sum_{i=1}^{n} (i) = \frac{n(n+1)}{2}$

(2) $\sum_{i=1}^{n} (i^2) = \frac{n(n+1)(2n+1)}{6}$

(3) $\sum_{i=1}^{n} (i^3) = \left(\frac{n(n+1)}{2}\right)^2$

So, if we partition [0,1] into n subintervals of size $(\Delta x_i) = \left(\frac{1}{n}\right)$

(each sub-interval is the base of one of the n rectangles), and let w_i be the x-value corresponding to the right-most point of each rectangle's base, then w_i will be ($\frac{i}{n}$), where i = 1,2, . . .,n. The situation with 10 rectangles is shown in Figure 11.

<u>Figure 11</u>

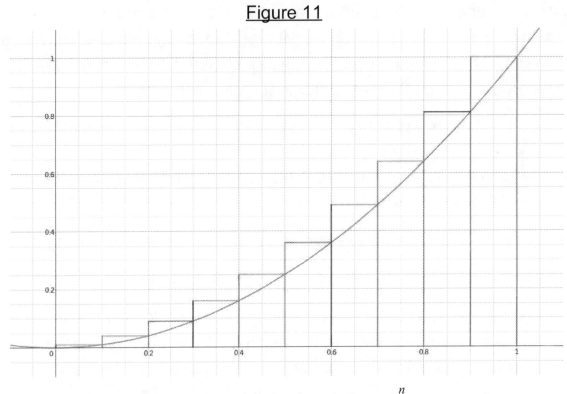

The area we are trying to find is approximated by $\sum\limits_{i=1}^{n} f(w_i)(\Delta x_i)$

$$= \sum\limits_{i=1}^{n} (w_i)^2 (\Delta x_i) = \sum\limits_{i=1}^{n} (\frac{i}{n})^2 (\frac{1}{n})$$

$$= (\frac{1}{n^3}) \sum\limits_{i=1}^{n} (i^2) = (\frac{1}{n^3}) \cdot \frac{n(n+1)(2n+1)}{6} = (\frac{1}{n^3}) \cdot \frac{2n^3 + 3n^2 + n}{6}$$

$$= (\frac{1}{3} + \frac{1}{2n} + \frac{1}{6n^2}).$$

So $\int\limits_{0}^{1} x^2 \, dx = \lim\limits_{n \to \infty} (\frac{1}{3} + \frac{1}{2n} + \frac{1}{6n^2}) = \frac{1}{3}$, since the last two terms in the sum tend to 0, as n → ∞. Obviously this only works well if f(x) is some kind of simple function. The method is very laborious compared to the simpler method that we will consider when we study the Fundamental

Theorem of calculus later in this chapter. We will consider one more example of using a Riemann sum to evaluate a definite integral.

Example 2: Find the $\int_{1}^{3} (9 - x^2)\, dx$ using Riemann sums.

The situation with 8 rectangles is shown in Figure 12.

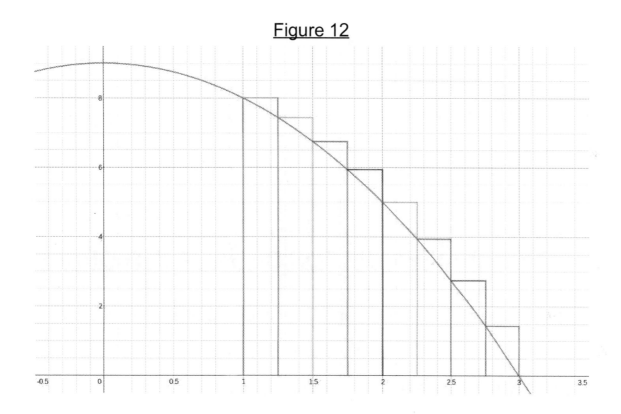

Figure 12

Now $(\Delta x_i) = \frac{b-a}{n} = \frac{3-1}{n} = \frac{2}{n}$. We will let w_i = the left-most point of the base of each rectangle. So $w_i = (1 + (i - 1)(\frac{2}{n}))$, (for i = 1,2,3, . . .,n).

So $\sum_{i=1}^{n} f(w_i)(\Delta x_i) = \sum_{i=1}^{n} (9 - (w_i)^2)(\Delta x_i) = \sum_{i=1}^{n} (9 - (1 + (i - 1)(\frac{2}{n}))^2)(\frac{2}{n})$

$= \sum_{i=1}^{n} [\, (\frac{-8}{n^3})\, i^2 + (\frac{16}{n^3} - \frac{8}{n^2})i - (\frac{8}{n^3} - \frac{8}{n^2} - \frac{16}{n})]$

$= 16 - (\frac{8}{n^2}) + (\frac{8}{n}) - (\frac{8}{n^3})(\frac{n(n+1)(2n+1)}{6}) + (\frac{16}{n^3} - \frac{8}{n^2})(\frac{n(n+1)}{2})$

$= (\frac{-8}{n^3})(\frac{2n^3+3n^2+n}{6}) + (\frac{12}{n}) + 12$

69

Therefore, as n $\rightarrow \infty$, the sum converges to $(12 - (\frac{8}{3})) = (28/3)$. This agrees with the shortcut methods which antiderivatives and the fundamental theorem of calculus provide for us.

Some properties of definite integrals:

(1) $\int_{a}^{a} f(x)\, dx = 0$.

(2) $\int_{a}^{b} f(x)\, dx = \int_{a}^{c} f(x)\, dx + \int_{c}^{b} f(x)\, dx$, if a,b, and c are values on an interval where f(x) is continuous.

(3) $\int_{a}^{b} f(x)\, dx = -\int_{b}^{a} f(x)\, dx$.

(25) THE MEAN VALUE FOR DEFINITE INTEGRALS

When we consider the case where f(x) is continuous on [a,b], then the function f(x) is bounded and $\int_{a}^{b} f(x)\, dx$ will be a finite number. So it is easy to see that there will exist a number M such that $\int_{a}^{b} f(x)\, dx = M \cdot (b-a)$. We call this result the mean value theorem for definite integrals, and M is the mean value.

(26) <u>THE FUNDAMENTAL THEOREM OF CALCULUS</u>

The evaluation of $\int_a^b f(x)\,dx$ with the limit of a Riemann Sum is a difficult way to evaluate the definite integral. In many cases it would not be possible at all. Fortunately, Isaac Newton and Wilhelm Leibniz in the 17th century, found a much simpler way. It involves antiderivatives. This result is so important, because it connects the two main parts of calculus which are known as the differential calculus and the integral calculus. This theorem has become known as the fundamental theorem of calculus.

<u>Statement and Proof of the Fundamental Theorem of Calculus:</u>

If f(x) is a continuous function on [a,b], and if $F(x) = \int_a^x f(t)\,dt$, then

   (1) F(x) is an antiderivative for f(x), and

   (2) If F(x) is any antiderivative for f(x), $\int_a^b f(x)\,dx = F(b) - F(a)$.

Proof:
(1) From the definition of the derivative (with $h = \Delta x$), we have

$$F'(x) = \lim_{h \to 0} \left(\tfrac{1}{h}\right)[F(x + h) - F(x)] = \lim_{h \to 0} \left(\tfrac{1}{h}\right)\left[\int_a^{x+h} f(t)\,dt - \int_a^x f(t)\,dt\right]$$

$$= \lim_{h \to 0} \left(\tfrac{1}{h}\right)\left[\int_x^a f(t)\,dt + \int_a^{x+h} f(t)\,dt\right] = \lim_{h \to 0} \left(\tfrac{1}{h}\right)\int_x^{x+h} f(x)\,dx .$$

From the mean value theorem for definite integrals, there exists a number M between x and (x + h) such that $F'(x) = \lim_{h \to 0} \left(\tfrac{1}{h}\right)Mh$

$= \lim_{h \to 0} (M) = f(x)$. M gets closer and closer to f(x) as h goes to 0.

(2) Let $F(x)$ and $G(x)$ be any two antiderivatives for $f(x)$ and c any number in $[a,b]$. Then if $F(x) = \int_{c}^{x} f(x)dx$,

$$\int_{a}^{b} f(x)dx = (\int_{a}^{c} f(x)dx + \int_{c}^{b} f(x)dx) = (\int_{c}^{b} f(x)dx - \int_{c}^{a} f(x)dx)$$

$$= F(b) - F(a).$$

For some real number d, $G(x) = F(x) + d$, since any two antiderivatives differ at most by a constant. So $G(b) - G(a) = (F(b) + d) - (F(a) + d) = F(b) - F(a)$.   QED

In the next section, we will see several examples of the use of the indefinite integral and definite integrals. The basic plan for evaluating a definite integral is to find an antiderivative $F(x)$ for the integrand $f(x)$ and then evaluate $F(x)$ between a and b. By evaluating it between a and b, we mean $(F(b) - F(a))$.

(27) **APPLICATIONS OF INTEGRATION**

Just as with the derivative, the definite integral and indefinite integration have many applications in modern science, mathematics, and engineering. We will provide some example applications: Finding the area under a curve, a couple physics examples, and two differential equations.

Example 1: Let $f(x) = 2x^4 + 3x^2 + 1$ on the interval $[1,3]$. We can calculate the area under $f(x)$ and above the x-axis on this interval

by evaluating $\int_{1}^{3} (2x^4 + 3x^2 + 1)\, dx = [\frac{2x^5}{5} + x^3 + x]_{1}^{3}$

$= ((\frac{2(3)^5}{5} + (3)^3 + 3) - (\frac{2(1)^5}{5} + (1)^3 + 1)) = (127.2 - 2.4) = (124.8)$.

Example 2: Suppose that a particle moves with speed s(t) = 3t³ + 4 during the time interval [1,5], where speed is measured in meters/sec and time in seconds. The total displacement in meters is $\int_{1}^{5}$ (3t³ + 4) dt

= [$\frac{3(t)^4}{4}$ + 4t $]_1^5$ = ((3(5)⁴/4 + 20) - (4.75)) = 484 meters.

Example 3: Suppose that a force acting on a particle in one dimension is given by f(x) = 3x on the interval [2,10], where the force is in units of Newtons and distance is measured in meters. In physics, the total work W (in units of Newton-meters - called Joules) done by f(x) on this displacement interval in the x-direction is given by

W = $\int_{2}^{10}$ f(x)dx = $\int_{2}^{10}$ 3x dx = [$\frac{3(x)^2}{2}$ $]_2^{10}$ = [(150) - (6)] = 144 Joules.

Example 4: The solution of differential equations involves indefinite integration to a significant extent. The subject of differential equations is a large and important area of mathematics, especially for the physical sciences and engineering. It is associated with other areas of mathematics such as linear algebra, Fourier analysis, Laplace transforms, infinite series, and other areas of advanced analysis. Sometimes we solve partial differential equations where the function involves more than one independent variable. In elementary algebra we solve equations for a number. In differential equations we are solving for a function like y = f(x) or z = f(x,y) from an equation involving the unknown function and its derivatives. Let's solve the differential equation $\frac{dy}{dx}$ = $\frac{4x^2-x+2}{e^y}$. This is called a variables separable differential equation. Multiplying through by e^y and dx yields:

$e^y dy = (4x^2 - x + 2)dx$. Taking the indefinite integral of both sides:

$\int e^y dy = \int (4x^2 - x + 2)dx$ leads to $e^y + C_1 = \frac{4x^3}{3} - \frac{x^2}{2} + 2x + C_2$. We

can combine the constants into one on the right-hand side and take the natural logarithm of both sides to get:

$y = \ln(\frac{4x^3}{3} - \frac{x^2}{2} + 2x + C)$. This is the solution. It can be verified by differentiating both sides. Note there are an infinite number of solutions for different values of the arbitrary constant C.

Example 5: Suppose we model the growth of a country's population, with the differential equation $\frac{dP}{dt} = kP(t)$.

This is a common way that exponential growth is modeled. What it says is that the rate of change of the population's growth is proportional to the size of the population. We have the two conditions P(0) = 200 million and P(10) = 255 million, where we suppose that time t = 0 (in years) corresponds to the beginning of 1990.

Multiplying through by dt and dividing by P(t) leads to $\frac{dP}{P} = kdt$. Integrating both sides we have, ln(P) = kt + C. Then exponentiating both sides gives $P(t) = e^{kt} \cdot e^C$, where e^C is a constant which we can just call C.

So P(t) = C e^{kt} is the solution of the differential equation.

From the first condition (at t = 0, P(t) = 200) we find that C = 200 million. From the second condition, $255 = (200)e^{10k}$. Then $\ln(\frac{255}{200}) = 10k$. So that k = .0243 = 2.43%. The constant k turns out to be the annual rate of growth, so the complete model of the country's population size at a time t in years is,

P(t) = 200 $e^{(.0243)t}$, in millions.

Assuming that the growth rate remains constant, we could expect a population of 367 million in the year 2015, which corresponds to t = 25.

At the end of the next chapter, we will mention how a measure of change in a physical system can be defined by a definite integral. In example 2, we therefore have measured a change in position (total distance traveled) by using a definite integral on an interval of time. In example 3, we have measured the amount of work done by using a definite integral on an interval of distance. A point we will be making is that change can be measured only over an interval of time or space, hence the importance of intervals in the measure of things.

(28) **FUNCTIONS OF TWO VARIABLES**

In this and the next section, to say the least, we will just barely touch upon the immense part of calculus dealing with functions of several variables. In the next two sections, we will discuss a little bit about partial differentiation and the double integral for a function of two variables.

A function of two variables can be written z = f(x,y), where x and y are the independent variables and z is the dependent variable. In applications of probability (a big part of this book) this is a surface above a region of the x-y plane. Let's assume that we are interested in a function f(x,y) defined on some open or closed region R in the x-y plane, and where f(x,y) is continuous on R. The idea of an open or closed region R in the plane is completely analogous to the idea of an open or closed interval on the real line. The idea of a continuous function f(x,y) on a region R of the plane is completely analogous to the idea of a continuous function f(x) on an interval of the real line. Continuity on a region R means that the $\lim\limits_{(x,y)\to(a,b)} f(x,y) = f(a,b)$ for all points (a,b) in R. The only difference is that this limit must be the same for all of the infinite number of ways that we can approach the point (a,b) in the plane, as compared to only two ways that we can approach a point on the real line. When we consider z = f(x,y), the derivative concept is generalized with so-called partial derivatives $\frac{\partial z}{\partial x}$ and $\frac{\partial z}{\partial y}$.

We say "the partial derivative with respect to x," and "the partial derivative with respect to y." These are defined at a point (a,b) in R in the following way:

$$\frac{\partial z}{\partial x}(a,b) = \lim\limits_{h \to 0} \left(\frac{f(a+h, b) - f(a,b)}{h} \right)$$

This tells us how f(x,y) is changing in the x-direction at the point (a,b,f(a,b)) when y is held constant at y = b.

$$\frac{\partial z}{\partial y}(a,b) = \lim\limits_{h \to 0} \left(\frac{f(a, b+h) - f(a,b)}{h} \right)$$

This tells us how f(x,y) is changing in the y-direction at the point (a,b,f(a,b)) when x is held constant at x = a.

As in the single variable case, there are easier ways of computing these partial derivatives for a given function f(x,y). The techniques of differentiation are the same, except that we simply treat all other variables as if they were constants. When differentiating f(x,y) with respect to x, we treat y as if it were a constant. When differentiating f(x,y) with respect to y, we treat x as if it were a constant.

<u>Example 1</u>: Find the partial derivatives $\frac{\partial z}{\partial x}$ and $\frac{\partial z}{\partial y}$ for

$z = f(x,y) = e^{xy} + 3x^3y^2 - 4xy$.

$\frac{\partial z}{\partial x} = ye^{xy} + 9x^2y^2 - 4y$ and $\frac{\partial z}{\partial y} = xe^{xy} + 6x^3y - 4x$.

Example 2: Find the partial derivatives $\frac{\partial z}{\partial x}$ and $\frac{\partial z}{\partial y}$ for
$z = f(x,y) = \cos(x^2 + 4xy) - \ln(2x + 3y)$.

$\frac{\partial z}{\partial x} = -\sin(x^2 + 4xy) \cdot (2x + 4y) - \frac{2}{(2x+3y)}$ and
$\frac{\partial z}{\partial y} = -\sin(x^2 + 4xy) \cdot (4x) - \frac{3}{(2x+3y)}$.

It should be clear that these techniques can be extended to functions of more than two variables. When we consider applied topics concerning functions of more than one variable, we usually have to consider vectors in two or three dimensions, and in addition we have to consider analytic geometry in the space of 3 dimensions. Vector analysis plays a big role in the theory of functions of several variables. There are so many topics that we are not considering here in this book. For a good understanding of the elementary concepts of probability and statistics, which is a major theme of this book, we will not need any more than the calculus of a single variable, which is what we have been concentrating on in this chapter.

(29) DOUBLE INTEGRALS

We have definite integration with a function of two variables f(x,y), analogous to the topic in the one variable case. In computing a double integral (an iterated double integral) we will use so-called multiple antidifferentiation. Just as with differentiation, the methods of antidifferentiation are exactly the same as in the single variable case, except that when working with more than one variable we have to treat all variables other than the one under consideration as if they are constants.

The theory behind definite integration of a continuous function f(x,y) over a region R of the x-y plane is completely analogous to definite integration of a function f(x) over an interval of the real line. In the single variable case, we break an interval [a,b] of the x-axis into a collection of equal sub-intervals $\{\Delta x_i\}$ and consider rectangles of height f(w_i), where (w_i) is any point in the sub-interval Δx_i , and f(x) is a continuous function on [a,b]. The limit of the sum of the rectangular areas as the number of rectangles goes to infinity gives us the exact area under f(x) on the interval [a,b]. In the two variable case, with equally spaced lines parallel to the x and y axes, we break the region R into a sub-collection of inscribed or circumscribed squares and consider a collection of n rectangular solids with base in the x-y plane and height f(u_i, v_i) , where (u_i, v_i) is any point in the square base. The volume of the collection of rectangular solids approximates the volume under f(x,y) and over R. Then we have a two-dimensional Riemann sum which will converge to the exact volume as the number of rectangular solids n goes to infinity. In symbols:

The $\displaystyle\lim_{n\to\infty} \sum_{i=1}^{n} f(u_i, v_i)\Delta x_i \Delta y_i \;=\; \iint_R f(x,y)\,dy\,dx$.

Example 1: Let R = {(x,y)| $1 \le x \le 3,\ 2 \le y \le 3$ }. Find the

$\iint_R f(x,y)\,dy\,dx$ when f(x,y) = $x^2 y^3 + 4$.

$$\iint\limits_R f(x,y)\,dy\,dx \;=\; \int\limits_1^3\int\limits_2^3 (x^2y^3+4)\,dy\,dx \;=\; \int\limits_1^3 [\tfrac{x^2y^4}{4}+4y]_2^3\,dx$$

$$=\;\int\limits_1^3 [(\tfrac{81(x^2)}{4}+12)-(4x^2+8)]\,dx \;=\; \int\limits_1^3 (\tfrac{65(x^2)}{4}+4)\,dx$$

$$=\;[\tfrac{65x^3}{12}+4x]_1^3 \;=\; (\tfrac{585}{4}+12)-(\tfrac{65}{12}+4) \;\approx\; 148.833.$$

Example 2: Find $\iint\limits_R (6x+10y)\,dy\,dx$,

where R = {(x,y)| $0\le x\le 10,\; x\le y\le 3x$} .

$$\iint\limits_R (6x+10y)\,dy\,dx \;=\; \int\limits_0^{10}\int\limits_x^{3x} (6x+10y)\,dy\,dx \;=\; \int\limits_0^{10} [6xy+5y^2]_x^{3x}\,dx$$

$$=\;\int\limits_0^{10} (52x^2)\,dx \;=\; (\tfrac{52}{3})[x^3]_0^{10} \;=\; \tfrac{52000}{3} \;\approx\; 17,333.33 \;.$$

In the study of functions of several variables we can have triple integrals, and even n-fold integrals (where n > 3) over spaces of higher dimension. We can use coordinate systems other than rectangular, such as polar, cylindrical, and spherical coordinate systems. Once again we have just barely scratched the surface of the calculus of several variables in the last two sections.

CHAPTER 5:
THE MEASURE OF NUMBERS

(30) **MEASURE**

This chapter addresses the measure of real number sets. The idea of measure is to quantify the amount of something in a set, such as the number of objects in the set, the length, area, or volume of the set, or the probability of the set, and so on. In this chapter we will show how we measure certain subsets of the real numbers, such as certain point sets and intervals. We will concentrate on discussing the measure of the rationals, irrationals, and the real numbers. Then we will discuss, based on the discussion of the measure of real number sets, my own speculations on the nature of physical reality regarding continuums of space and time, which are models based on the real numbers.

(31) **A DEFINITION OF MEASURE**

A measure 'm' of real sets satisfies the following two conditions:

(1) If A is a real number set, the measure of A, denoted m(A), is always non-negative. That is $m(A) \geq 0$.

(2) If we have a countable collection {A$_i$} of disjoint sets, then

$$m\left(\bigcup_{i=1}^{\infty} A_i\right) = m(A_1) + m(A_2) + \cdots$$

(32) <u>THE MEASURE OF REAL NUMBER SETS</u>

A real interval is any continuous subset of the real numbers of the
following types: an open interval (a,b), a closed interval [a,b], or a
half-open/half-closed interval (a,b] or [a,b), where a < b in each case.
The measure of any one of the above real intervals is defined to be
(b-a). We can call this type of measure Euclidean measure. Denote
the collection of all real intervals of the above four types together as
$\{\Delta(a,b) \mid a, b \in \mathbb{R}, \ a < b\}$. Note that the measure of any interval
$\Delta(a,b)$ is $m(\Delta(a,b))$ = (b - a) which is always greater than 0.

<u>A Measure Space</u>
We want to find the measure of any real number set X which is a subset
of $\mathbb{R}$. In this book, which deals mainly with the real numbers, the
measure that we are interested in is the Euclidean measure of the real
number set X. To find the measure of any real number subset X, it is
sufficient to consider only a subcollection of all the possible subsets of
$\mathbb{R}$, a subcollection of sets that are positively measured. The subsets of
$\mathbb{R}$ that we choose are the set of all intervals $\{\Delta(a,b) \mid a, b \in \mathbb{R}, \ a < b\}$.
We call this subcollection of positively measured sets $\sum$. The idea is
that we can use the measure of sets from $\sum$ to find the measure of the
set X. The triple $(\mathbb{R}, \sum, m)$ is called a measure space.

<u>Convergence of Intervals and Convergence of Measure</u>
We are interested in the measure of some set of real numbers X,
either an interval or a set of points. If we have a sequence of intervals

82

$\{\Delta(a_n, b_n)\}$ such that X = $\lim\limits_{n \to \infty} \Delta(a_n, b_n)$, then we say that we have

convergence of a sequence of intervals to the set X. If this is the case, then it is clear that the m(X) = $\lim\limits_{n \to \infty} m(\Delta(a_n, b_n))$. We then say that we

have convergence of measure.

The Measure of a Point

The first thing that we will prove is that the measure of a point is 0. Let X = {a} be a set consisting of a single real number 'a.' Let $\{\Delta(a_n, b_n)\}$ = $\{[a - \frac{1}{n}, a + \frac{1}{n}]\}$. Then each interval in the sequence contains a, and the sequence of intervals converges to the set {a} consisting of the real number a. The m(X) = $\lim\limits_{n \to \infty} m([a - \frac{1}{n}, a + \frac{1}{n}])$ = $\lim\limits_{n \to \infty} (a + \frac{1}{n} - (a - \frac{1}{n}))$

= $\lim\limits_{n \to \infty} (\frac{2}{n}) = 0$. Therefore the measure of any point is 0.

The Measure of a Countably Infinite Set of Points

Let $\{c_i\}$ be a countably infinite set of real numbers (points).

Let $\varepsilon > 0$. Then let $N_\varepsilon(c_i)$ = $(c_i - \varepsilon(\frac{1}{2})^i, c_i + \varepsilon(\frac{1}{2})^i)$ be an open cover of

the real number c_i. Then the m($N_\varepsilon(c_i)$) = $2\varepsilon(\frac{1}{2})^i$ (for i = 1,2,3, . . .).

The measure of $\{c_i\}$ < $\sum\limits_{i=1}^{\infty} 2\varepsilon(\frac{1}{2})^i$ = 2ε, since we have a geometric series

and there is overlapping of our open cover of the elements of $\{c_i\}$. Letting $\varepsilon \to 0$, we see that the measure of the countably infinite set of real numbers $\{c_i\}$ = 0.

From this result, it is clear that the measure of the Rationals is 0. The real line, if it consisted only of the Rationals, would be totally incomplete. It would not be a continuum. As a corollary to this, the measure of the Integers is also 0, since they are a subset of the rationals.

The Measure of $\mathbb{R}$

If we break any interval $\Delta(a, b)$ into an infinite number of disjoint positively measured sub-intervals $\{\Delta(a_n, b_n)\}$, then the measure of the union of the sub-intervals is $\sum_{n=1}^{\infty} m(\Delta(a_n, b_n)) = m(\Delta(a, b)) = (b-a)$. We claim that this collection of sub-intervals must be countably infinite in number.

Proof: Each of the sub-intervals has measure > 0, which means that each sub-interval $\Delta(a_n, b_n)$ is an uncountably infinite continuum within $\Delta(a, b)$. Consider any sub-interval in the sequence $\Delta(a_n, b_n)$, and choose c_n to be any rational number such that $a_n < c_n < b_n$. Then the set $\{c_n\}$ is a subset of the rationals in $\Delta(a, b)$. Therefore they are countably infinite in number. Therefore the collection of sub-intervals is countably infinite in number. $\qquad$ QED

Note this same argument works if we let a = $-\infty$, and b = ∞. In other words if we have an infinite collection of positively measured and disjoint real intervals $\{\Delta(a_n, b_n)\}$ whose union is $\mathbb{R}$, then this collection of sets is countably infinite in number. One way that we can write $\mathbb{R}$ as a countable union of disjoint positively measured sets is to let $\mathbb{R}$ be the union of the real intervals $\{$ [k,k+1) | k $\in \mathbb{Z}\}$, then it is clear that

$$m(\mathbb{R}) = \sum_{k \in \mathbb{Z}} m[k,k+1) = \infty .$$

Measure of the Irrationals

If we have a closed interval of real numbers X = [a,b], where a < b, m(X) = (b-a). The set X consists of the rationals in X plus the set of irrationals in X, which we can write $X = \mathbb{Q}(X) \cup Irr(X)$. Since $\mathbb{Q}(X)$ and Irr(X) are disjoint sets, m(X) = $m(\mathbb{Q}(X))$ + $m(Irr(X))$ = 0 + $m(Irr(X))$. So the m(X) = m(Irr(X)). Thus the irrational numbers in any real interval [a,b] are uncountably infinite and have measure (b-a) also. This says that far and away, most real numbers are irrational numbers.

(33) <u>NECESSITY OF CONSIDERING INTERVALS</u>

Consider the real interval [0,1]. We wish to know its measure. We can consider the uncountably infinite set of real numbers in this interval and include each number as a set A_x in the set Σ. A_x contains a single point x, where x is in [0,1]. Then since the measure of a real number is zero, the measure of the union of all the disjoint sets A_x is the sum of an uncountably infinite number of zeros, which would equal zero! This is contrary to the common sense notion of Euclidean measure that the measure of the real interval [0,1] is (1-0) = 1. It would follow that any interval of real numbers would have measure 0, and the measure of $\mathbb{R}$ itself would be 0. This is unsatisfactory! So Σ must contain all of the real number intervals with positive measure and need not contain any points. We need to think in terms of intervals instead of points.

(34) <u>INTERVALS AND CONTINUUMS</u>

Instead of thinking of the real numbers and physical concepts, like space and time, as being made up from points, let's think of them as being made up of arbitrarily small real intervals $\{\Delta(a_i, b_i)\}$, where the $m(\Delta(a_i, b_i)) > 0$. We can call these very small real intervals continuums (Δc_i), and in any real interval X there can be an infinite number of them. When just considering all of the continuums in X, there would be an uncountably infinite number of them, but if X is the union of an infinite number of disjoint continuums, then there would only be a countably infinite number of them.

$\Delta(a, b)$ is any of the four types of intervals that we described above: open, closed, or the two types of half-open/half-closed intervals. Within $\Delta(a, b)$, think of a continuum Δc_i as being any one of these four types of intervals also, with measure > 0. Then the interval $\Delta(a, b)$ can be a countably infinite union of these disjoint positively measured continuums. So if the real number interval $\Delta(a, b)$ is an infinite union of disjoint positively measured continuums $\{\Delta c_i\}$, then $\Delta(a, b) = \bigcup_{i=1}^{\infty}(\Delta c_i)$ and the m($\Delta(a, b)$) = $\sum_{i=1}^{\infty}$ m(Δc_i) is a convergent infinite series with sum (b-a).

From above we know that if $\mathbb{R}$ is the union of a countably infinite number of disjoint continuums $\{\Delta c_i\}$, each of positive measure, then we write $\mathbb{R} = \bigcup_{i=1}^{\infty}(\Delta c_i)$, and m($\mathbb{R}$) = $\sum_{i=1}^{n} m(\Delta c_i)$ = ∞. My point here is to think of continuums, not points, as what the real numbers and physical continuums are actually made of. This says that continuums are real, and points are not.

(35) MATHEMATICAL REALITY AND PHYSICAL REALITY

Certainly, the use of the term point to denote a single real number doesn't seem to be problematic, but as an actual physical object, we must question whether a point can be said to exist. It has no extension in any of the three mutually perpendicular directions that we think of when we think of space. If you ask somebody to take a pencil and a sheet of paper and draw a point, they would usually draw a small point on the paper, like writing a period. However, if we were to magnify this it would look nothing like a point. The best way that a point could be represented on a sheet of paper would be to hold the pencil above the paper and then remove the pencil, leaving no mark on the paper at all. Since this is the case, we have to ask once again does a point as a physical object really exist? I think that points have no reality in the physical world. I think that a number as a mathematical symbol has reality, but of course this is different from physical reality. Use of the word point to represent a number, or a position on the real line, or as a location in the coordinate plane, or as a location in a three dimensional coordinate system has mathematical reality. We can call this mathematical reality, but without physical reality.

Returning to pencil and paper again, we can ask somebody to draw a line segment of length one inch on the paper. They will draw a line segment of some length (where the accuracy doesn't matter here). One could say that the points of a geometric line segment, and hence the geometric line segment itself also has no reality, because the line segment is made up of an uncountably infinite number of points, each of which has no extension in any of the three spatial dimensions. So we have a problem here! I think that the way out of this predicament is to say (as like an axiom) that any mathematical or physical object that supposedly has extension greater than zero in at least one of the three dimensions has physical reality. I say this because we deal with objects everyday that obviously have some kind of non-zero spatial extension and we don't deal with anything that doesn't have any extension at all in any of the three dimensions. So I say that any mathematical object, including line segments, lines, curves, planes, surfaces, and volumes that have some kind of spatial extension are real and can correspond to things that have physical reality.

I think that when we deal with real number intervals of positive extension, we move into a situation where we are forced to ignore that a segment consists of points, but rather consider only that it is a segment (a continuum) and nothing more. My point is that the only things existing in physical reality are objects that have extension. The extension would be like a line segment, area, or volume represented with a real number that is greater than zero. Forget all about that they are supposedly made up of points! In geometry textbooks, we read of the undefined terms point, line and plane, and these things are the starting point for the mathematics to follow in those textbooks. The use of points, with the real numbers, provides a very good approximation to the physical worlds of space and time. However, I say that for the physical world, the undefined term should be something like a linear continuum or linear continuums, which can be arranged orthogonally to each other. I believe that the whole of the geometry of physical space could be constructed from these continuums.

Instead of everything being made of points, everything is made up of continuums, no matter how small. The important thing about continuums is simply that they all have length greater than zero. This may make it sound that the physical world, based on intervals, takes on some kind of a mysterious or spiritual quality, but we have to take into consideration that points, and finite or countably infinite collections of them, have measure 0. The number zero is indispensible in mathematics as a symbol and for its use in our number systems, but objects of zero extent do not correspond to anything in the physical world.

(36) <u>SPACE, TIME, AND CHANGE</u>

The continuum of the real numbers is a model that we have made (or discovered) for the continuums of space and time. It has been a central theme so far in this book that the continuum of the real line contains an uncountably infinite collection of numbers. We have stressed that individual points on the real line have measure zero, and therefore this means that we have to question the physical reality of points in space and time.

I have hypothesized that the starting point for geometric considerations in the physical world should be continuums rather than points. Points are the building blocks of the mathematical world, continuums are the building blocks of the physical world. Things of non-zero measure are the only things which have a physical reality because things in the physical world are only real if they have some non-zero extension in at least one dimension. I think that continuums may have a reality without regard to the points that they supposedly contain. Continuums such as with real number intervals, and as with intervals of space and time are the true starting point for the measure of things.

With regard to time it would follow that the present moment does not exist. There is only some intangible boundary between two great continuums, the past and the future, and it is constantly changing. We cannot put our finger on it because it does not exist. When we measure the length of some object, it is always greater than zero. We cannot measure with any kind of measuring device the length of things with zero extension because such objects do not exist in the first place!

Time is dependent on how much the collection of material objects in the universe are changing position. When we look briefly at a clock, we don't in principle see a change occurring. We only notice a change in the position of the hands of the clock when we take our eyes off of it and look at it again at a later time. If every particle in the universe suddenly stopped moving, then time would cease to exist, since time is only measured by the change in position of material objects, and in particular some mechanical device with moving parts that we call a clock. If all motion stopped everywhere, including the parts of the clock, then there would be no concept of time.

An observer can take only one measurement of a state of some part of the physical universe at any given point of time t_1. A subsequent measurement would be at a time t_2, where $(t_2 - t_1) = \Delta t > 0$. So in a given interval of time $\Delta(a, b)$, there is at most a countable number of measurements and hence at most a countable number of disjoint positively measured sub-intervals of time making up $\Delta(a, b)$. When we make a measurement at each of the times $\{t_i\}$, we are measuring a change of state of some physical system in the universe, since the last measurement. Since we and our machines can only make a countable number of measurements in an interval of time, because there are limitations on the speed of material objects through space and on light itself (which means that it takes a positive amount of time for signals to travel between different regions of space), we see the world in a discrete way, a sequence of discrete steps. This is an observer based limitation that we have in measuring a change in a physical system. We are constrained so as not to be able to witness continuous change. Perhaps this is an argument for the primacy of continuums in the mathematical and the physical world.

It seems that there is a problem for objects traveling through a space and time consisting of an uncountable number of points, because for any real number we cannot name the next point. There is no adjacency of points within the real numbers. It's sort of like the fact that we cannot drive a car from Germany to Spain without first driving through France. We need these adjacent regions. Perhaps if the reality of space and time is that they are actually made up of continuums, then we would have the adjacent regions that we need in order to have motion from one part of space to another.

Zeno was a thinker from the ancient world that proposed many logical paradoxes concerning motion. One way of describing (in modern language) the essence of many of his paradoxes is in the following way: Suppose we have something traveling from point a to b. Zeno said that the something must travel the first half of the way from a to b, then half of the remaining distance that remains, and then half of the still remaining distance, and so on ad infinitum. What Zeno is trying to say is that the something should never make it from a to b because it must make an infinite number of "half of the remaining distance" journeys! This is a logical problem that has confounded many thinkers from antiquity to the present, even though it may seem a little bit silly. I think it clearly shows how logic cannot always be trusted in mathematical considerations involving infinity. The logic that is used in the first chapters of this book dealing with set theory and in the next chapters concerning probability is just the logic of "or," "and," and "not". These parts of logic which are sort of built into set theory seem to be very solid concepts.

We know from physics that distance "d" equals velocity "v" multiplied by a time "t", or d = vt. Suppose velocity v for some mass moving in space (which quantifies somehow a measure of energy imparted to it at an earlier time) is considered constant for some small intervals of time Δt and space $\Delta(a, b)$. We can say that the measure of the interval of space is m($\Delta(a, b)$) = $\sum_{i=1}^{\infty} m(\Delta c_i)$, and the measure of the interval of time is $m(\Delta t) = \sum_{i=1}^{\infty} m(\Delta t_i)$ (we are saying that the intervals of space and time are a union of disjoint continuums). Then since t = $\frac{d}{v}$, we could say that the time needed to traverse the space interval is (in terms of sums) $\sum_{i=1}^{\infty} m(\Delta t_i) = (\frac{1}{v}) \cdot \sum_{i=1}^{\infty} m(\Delta c_i)$. We know that the sum of the measures of the contiguous space continuums converges to (b-a) < ∞, so therefore the space interval is traversed in the finite time given above by $m(\Delta t) = \sum_{i=1}^{\infty} m(\Delta t_i) = (\frac{1}{v}) \cdot \sum_{i=1}^{\infty} m(\Delta c_i)$. What does this have to say about

Zeno's Paradox? and the application of the logic from Zeno's Paradox to mathematical and physical problems?

Physicists have postulated the existence of something called the Planck length (about 1.6×10^{-35} meters), which is supposedly the smallest possible measureable length in the physical world. This seems to remind me of one of the older formulations of Calculus known as the Infinitesimal Calculus, which has been largely supplanted in modern times by a standard formulation where it is assumed that intervals of distance or intervals of time, etc. are infinitely divisible, that there is no need to assume a smallest length that is positive in measure. Certainly this makes sense mathematically, but one may wonder if there is a smallest measureable length in the physical world. Maybe the idea of continuums is not far from the truth. This very probably is a concept that I am not the first to consider - I am not an expert on the history of such considerations and hypotheses. If there is a smallest measure of space and time and maybe other quantities in the physical world, then it seems that many of my summations in this chapter would turn out to be finite summations! Did God make the world, space as well as material objects, out of a collection of very small pieces?

Earlier in the book, we discussed integral calculus, and realized that we can have a "measure of change in position" or a "measure of total work done" that is calculated with a definite integral. These measures of change can sometimes be negative and sometimes positive. For example, when a particle is in motion along a line during a certain interval of time, its total displacement can be negative or positive. We call such a measure which can take positive or negative values a signed measure. With definite integrals representing some kind of change, notice that a definite integral is only non-zero when integrating over an interval, again emphasizing the importance of intervals when measuring something.

CHAPTER 6:
INTRODUCTION TO PROBABILITY

(37) **PROBABILITY SPACES**

When we conduct a random experiment, the set of all possible outcomes is called the sample space $\mathbb{S}$. A random experiment is one where we cannot predict what the outcome will be beforehand. In this treatment of probability and statistics, the members of $\mathbb{S}$ are real numbers. Sets that we assign a probability to are called events, and the set of all events is called the event space $\mathbb{E}$. The event space $\mathbb{E}$ is a collection of positively measured subsets of the sample space $\mathbb{S}$, from which the probability measure of any subset X of $\mathbb{S}$ can be determined. A probability measure is a function Pr(x), such that:

(1) $Pr(A) \geq 0$, for all events A in $\mathbb{E}$.
(2) If $\{A_1, A_2, \ldots\}$ are disjoint events in $\mathbb{E}$, then

$$Pr \left(\bigcup_{i=1}^{\infty} A_i \right) = Pr(A_1) + Pr(A_2) + \cdots$$

(3) $Pr(\mathbb{S}) = 1$.

The triple $(\mathbb{S}, \mathbb{E}, Pr)$ is called a probability space. Note that a probability space is a particular type of measure space. With the measure space $(\mathbb{R}, \Sigma, m)$ of the real numbers, $m(\mathbb{R}) = \infty$. With the measure space $(\mathbb{S}, \mathbb{E}, Pr)$ associated with some random experiment, the $Pr(\mathbb{S}) = 1$. A probability space defines a probability distribution X. When $\mathbb{S}$ is a finite or countably infinite set, we have a discrete distribution. When $\mathbb{S}$ is an uncountably infinite set, we have a continuous distribution. When we have a discrete distribution, the measures of individual numbers in $\mathbb{S}$ are non-negative probabilities and the probability of events is computed using summations. If we have a continuous distribution, then the measure (probability) of any single number is 0, just like the measure of any single number on the real line. For continuous distributions, only intervals have non-zero probability and are computed using integrals.

Names of distributions are capital letters, while we denote particular outcomes with lower-case letters. There is no universally agreed upon definition of probability, so it has to be done axiomatically like above. In this chapter, we will use the counting measure.

A probability is a number in the interval [0,1]. An event with probability 0 is an impossible event. An event with probability 1 is an event certain to occur. Usually if A is an event of interest, then the usual case is that $0 < \Pr(A) < 1$. If we have two events A and B with $0 < \Pr(A) < \Pr(B) < 1$, we would say that event A is less likely to occur than event B, or that event B is more likely to occur than event A. With discrete distributions the sum of the probabilities of all the outcomes in the sample space is always 1. With continuous distributions the integral over the entire sample space is always 1. When multiplying a probability by 100%, we are expressing a probability as a percentage in the interval [0%, 100%]. These are the two ways that we express a probability.

(38) <u>COUNTING MEASURE AND PROBABILITY</u>

Suppose we have an experiment that has N possible outcomes, all of which are equally-likely to occur. Then $\mathbb{S} = \{x_1, x_2, \ldots, x_N\}$ is the sample space. For a finite sample space $\mathbb{S}$, the event space $\mathbb{E}$ would be all subsets of $\mathbb{S}$, which includes single outcomes, or any other collection of the outcomes in $\mathbb{S}$, including $\oslash$ and $\mathbb{S}$ itself. There are 2^N possible subsets of $\mathbb{S}$ in this case. If event A is a single outcome, the $\Pr(A) = \frac{1}{N}$. So all of the outcomes in $\mathbb{S}$ are equally-likely, which leads to the counting measure for the probability of an event E in the event space $\mathbb{E}$:

<u>Counting Measure for Probability</u>:

If E $\in$ $\mathbb{E}$, then: Pr(E) = ($\frac{card(E)}{card(sample\ space)}$).

For example, if we have some experiment with an equally-likely sample space $\mathbb{S}$ = {0,1,2,3,4,5,6}, and event E = {1,3}, and event F = {0,2,4},

Then the Pr(E) = ($\frac{card(E)}{N}$) = 2/7, and the Pr(F) = ($\frac{card(F)}{N}$) = 3/7.

This means that if we repeated the experiment a large number of times, the event E would occur about 2 out of every 7 repetitions, and event F would occur about 3 out of every 7 repetitions.

(39) **PROBABILITY OF COMPOUND EVENTS**

The following rules apply for all possible probability measures, but we will explain them assuming that all outcomes in $\mathbb{S}$ are equally-likely. The main topics are the sum rules, conditional probability, the product rules, and Bayes Theorem. But first, let me mention the notion of odds:

In a situation where we have equally-likely outcomes, we write the odds in favor of an event E to be (a:b) (pronounced 'a' to 'b') if there are (a+b) total outcomes in S, and a of them are favorable to E and b are favorable to $\widehat{E}$. The odds in favor of $\widehat{E}$ are (b:a). If given that the odds in favor of event E are (a:b), then the Pr(E) = $\frac{a}{a+b}$. Similarly, the Pr($\widehat{E}$) = $\frac{b}{a+b}$. The Pr(E) + Pr($\widehat{E}$) = 1, since $\frac{a}{a+b}$ + $\frac{b}{a+b}$ = 1. Here events

E and $\widehat{E}$ are disjoint (meaning that they have no overlap in the sample space) and they're exhaustive events (since E $\cup \widehat{E}$) = S). Whenever we have two disjoint and exhaustive events A and B, the Pr(A) + Pr(B) = 1.

Sum Rules: Suppose we have a sample space S = {-3, -2, -1, 0, 1, 2, 3, 4, 5, 6, 7}, and all outcomes are equally-likely. Define the events A = {-2, -1, 0, 1}, B = {3, 4, 5}, and C = {0, 1, 6, 7}.

From our definition of probability, the Pr(A) = ($\frac{card(A)}{N}$) = 4/11, the Pr(B) = ($\frac{card(B)}{N}$) = 3/11, and the Pr(C) = ($\frac{card(C)}{N}$) = 4/11.

Since A and B are disjoint, Pr(A $\cup$ B) = Pr(A) + Pr(B) = 4/11 + 3/11 = 7/11.

If we tried to compute the Pr(A $\cup$ C) the same way that we computed Pr(A $\cup$ B), the Pr(A $\cap$ C) would be double-counted since A and C are not disjoint. The Pr(A $\cap$ C) is equal to the Pr({0, 1}) = $\frac{2}{11}$, and it is included in Pr(A) and Pr(C), so it needs to be subtracted once.

The Pr(A $\cup$ C) = Pr(A) + Pr(C) - Pr(A $\cap$ C) = $\frac{4}{11}$ + $\frac{4}{11}$ - $\frac{2}{11}$ = $\frac{6}{11}$.

Thus we have a general sum rule for probability:
Pr(E $\cup$ F) = Pr(E) + Pr(F) - Pr(E $\cap$ F), for any two events E and F.

This can be generalized to more than two events. We will consider the case where we have three events E, F, and G:

We start by writing Pr(E $\cup$ F $\cup$ G) = Pr(E) + Pr(F) + Pr(G). The Pr(E $\cap$ F), Pr(E $\cap$ G), and Pr(F $\cap$ G) have been double-counted, and the Pr(E $\cap$ F $\cap$ G) has been triple-counted.

So then write Pr(E ∪ F ∪ G) = Pr(E) + Pr(F) + Pr(G)
$\qquad$ - Pr(E ∩F) - Pr(E ∩ G) - Pr(F ∩ G).

Now The Pr(E ∩F ∩ G) has been added and subtracted three times, so we need to add it once.

Thus we have the general sum rule for three events:
The Pr(E ∪ F ∪ G) = Pr(E) + Pr(F) + Pr(G)
$\qquad$ - Pr(E ∩F) - Pr(E ∩ G) - Pr(F ∩ G) + Pr(E ∩F ∩G)

The special case for disjoint events E and F (where (E ∩ F) = ∅) is:
Pr(E ∪ F) = Pr(E) + Pr(F).

This can be generalized to any number of disjoint events:
For three disjoint events E, F, and G,
Pr(E ∪ F ∪ G) = Pr(E) + Pr(F) + Pr(G)

<u>Conditional Probability</u>: Suppose, associated with some experiment, we have the sample space $\mathbb{S}$ = {-3, -2, -1, 0, 1, 2, 3}, where the outcomes are equally-likely.

Let A be the event {-1, 0, 1, 2, 3} and B the event {2, 3}.
Then (A ∩B) = {2, 3}.

We define the Pr(B|A), stated "The probability of B given that A occurs," or simply "The probability of B given A," to be

Pr(B|A) = $\frac{Pr(A ∩ B)}{Pr(A)}$.

In this case, the $\Pr(B|A) = \frac{Pr(A \cap B)}{Pr(A)} = \frac{(2/7)}{(5/7)} = (\frac{2}{5})$. Note that this is like restricting the sample space to the event A.

Product Rules: We can re-write the conditional probability formula to get the so-called general product rule for any two events A and B:

$$\Pr(A \cap B) = \Pr(A) \cdot \Pr(B|A) \quad \text{or} \quad \Pr(A \cap B) = \Pr(B) \cdot \Pr(A|B)$$

We have to use this conditional probability notation since events A and B are not necessarily independent. For example, if we are drawing two cards from a well-shuffled deck, without returning the first card to the deck before drawing the second (sampling without replacement), and define the events:

A is the event that a red card is drawn on the first draw.
B is the event that a red card is drawn on the second draw.

Then $\Pr(A) = \frac{26}{52}$ and the $\Pr(B|A) = \frac{25}{51}$. So the
$\Pr(A \cap B) = (\frac{26}{52}) \cdot (\frac{25}{51}) = (\frac{650}{2652}) \approx 0.2451$.

If we draw the two cards, returning the first card to the deck before drawing the second (sampling with replacement), and define events A and B in the same way,

Then the $\Pr(A) = \frac{26}{52}$ and the $\Pr(B|A) = \frac{26}{52}$. So the
$\Pr(A \cap B) = (\frac{26}{52}) \cdot (\frac{26}{52}) = (\frac{676}{2704}) = 0.25$.

The two results are not the same. In the first case A and B are dependent events. In the second case A and B are independent events. Any two events A and B are independent if $\Pr(A|B) = \Pr(A)$ or if $\Pr(B|A) = \Pr(B)$. These two formulations mean that knowing that B has occurred does not have any influence on the occurrence of A in the first case, and vice versa in the second case. The product rule can be extended to more than two events: If A, B, and C are any three events,

The $Pr(A \cap B \cap C) = Pr(A) \cdot \dfrac{Pr(A \cap B)}{Pr(A)} \cdot \dfrac{Pr(A \cap B \cap C)}{Pr(A \cap B)}$

$$= Pr(A) \cdot Pr(B|A) \cdot Pr(C|(A \cap B))$$

There is a special case of the product rule when the events are independent. Suppose we have, associated with some experiment, the sample space $\mathbb{S} = \{0, 1, 2, 3, 4, 5\}$, where the outcomes are equally likely. The experiment is repeated twice and the trials are independent, meaning that the outcome of the first trial has no influence on the second trial.

Let A be the event that the outcome of the first trial is in $\{0, 2\}$.
Let B be the event that the outcome of the second trial is in $\{1, 2, 3\}$.

Then the $Pr(A \cap B) = Pr(A) \cdot Pr(B) = (\frac{2}{6}) \cdot (\frac{3}{6}) = \frac{1}{6}$. So we have the product rule for independent events E and F:

$Pr(E \cap F) = Pr(E) \cdot Pr(F)$.

The above formula can easily be extended to more than two independent events $\{E_1, E_2, \ldots, E_n\}$:

$Pr(E_1 \cap E_2 \cap \cdots \cap E_n) = Pr(E_1) \cdot Pr(E_2) \cdots Pr(E_n)$.

From an experiment, a sample of n independent observations occurs, that is, $\{E_1, E_2, \ldots, E_n\}$ is collected. Each outcome is an event and the

$$Pr(E_1 \cap E_2 \cap \cdots \cap E_n) = Pr(E_1) \cdot Pr(E_2) \cdots Pr(E_n).$$

This is the probability of occurrence of that particular sample.

Bayes Theorem: This theorem is the basis for an alternative approach to statistical inference called Bayesian statistical inference. Most of statistical methodology, at least at the basic level is based on the relative frequency approach, and this is historically the most common approach. The subject of probability is of course the same in both approaches and fortunately the two usually produce very similar results. In the relative frequency approach, the parameters of a distribution are simply numbers which specify a particular distribution. We use data from experiments to make inferences about these fixed parameters. This is the approach that we use in this book.

In the Bayesian approach, a parameter is considered to have a distribution instead of being a fixed number. There is a prior assignment for the distribution of the parameter, called its prior distribution. Then the data collected in an experiment is used to update the prior distribution for a parameter to a posterior distribution for the parameter. So we have this initial guess, the prior distribution, for the parameter's distribution (which tells us where it is believed the parameter is most likely to be). In the relative frequency approach, we make no initial guesses about the parameters, and the data is all that we have to go on. Bayesian statisticians believe that we should use all information about a parameter that we may have before we conduct the experiment, and this is expressed in the prior distribution.

We will discuss here the most basic form of Bayes theorem, where the sample space for a parameter is broken up into a set of n mutually exclusive(disjoint) and exhaustive events $\{A_1, A_2, \ldots, A_n\}$. These events and their probabilities form the prior distribution for the $\{A_i\}$. The result of the experiment we will call the event E. We have knowledge of the conditional probabilities $\{Pr(E|A_1), Pr(E|A_2), \ldots, Pr(E|A_n)$, which we use to update the prior probabilities to the posterior distribution probabilities $\{Pr(A_1|E), Pr(A_2|E), \ldots, Pr(A_n|E)\}$ in the following way:

$$\text{For i = 1,2, \ldots,n,} \quad \text{the } Pr(A_i|E) = \frac{Pr(A_i \cap E)}{Pr(E)} = \frac{Pr(A_i)Pr(E|A_i)}{Pr(E)}$$

$$= \frac{Pr(E|A_i)Pr(A_i)}{Pr(E|A_1)Pr(A_1) + \cdots + Pr(E|A_n)Pr(A_n)} \cdot$$

(40) SOME PROBABILITY CALCULATIONS

Example 1: Suppose a company consists of 5 men and 9 women. A five member committee is to be selected from among the employees for a business trip. Every employee has an equal chance of being selected. What is the probability of the event E that 3 men and 2 women are selected?

There are C(5,3) ways that 3 men can be selected from the 5 and C(9,2) ways that 2 women can be selected from the 9. Then from the FPC, there are C(5,3)·C(9,2) ways that E can happen. There are a total of C(14,5) committees in the sample space.

$$C(5,3) \cdot C(9,2) = \tfrac{5!}{(2!)(3!)} \cdot \tfrac{9!}{(7!)(2!)} = 360. \quad C(14,5) = \tfrac{14!}{(9!(5!)} = 2002.$$

So the Pr(E) = $\frac{card(E)}{card(sample\ space)}$ = $\tfrac{360}{2002} \approx (0.18)$ or 18%.

Example 2: Suppose we have an experiment with sample space

$\mathbb{S} = \{0,1,2,3,4,5,6,7,8,9,10\}$, where the outcomes are equally-likely.

Let E be the event $\{0,1,4,7,10\}$.
Let F be the event $\{1,3,7,8,10\}$.
Let G be the event $\{3,5,7,9,10\}$.
$(E \cap F) = \{1,7,10\}$,
$(E \cap G) = \{7,10\}$,
$(F \cap G) = \{3,7,10\}$
$(E \cap F \cap G) = \{7,10\}$

We wish to know the $Pr(E \cup F)$ and $Pr(E \cup F \cup G)$?
$Pr(E \cup F) = Pr(E) + Pr(F) - Pr(E \cap F) = \frac{5}{11} + \frac{5}{11} - \frac{3}{11} = \frac{7}{11}$.

$Pr(E \cup F \cup G) = Pr(E) + Pr(F) + Pr(G)$
$\qquad\qquad - Pr(E \cap F) - Pr(E \cap G) - Pr(F \cap G) + Pr(E \cap F \cap G)$
$\qquad = \frac{5}{11} + \frac{5}{11} + \frac{5}{11} - \frac{3}{11} - \frac{2}{11} - \frac{3}{11} + \frac{2}{11} = \left(\frac{9}{11}\right)$.

Example 3: A license plate has 6 characters, the first 3 must be letters of the alphabet and the last three must be numeric digits 0 through 9. The plates are randomly produced. Define the event E to be the production of a license plate where each of the letters is different and each of the numbers is different.

The number of elements in E is $(26)(25)(24)(10)(9)(8) = 11,232,000$.
The total number of elements in $\mathbb{S}$ is $(26)^3(10)^3 = 17,576,000$.

Therefore the $Pr(E) = \dfrac{card(E)}{card(sample\ space)} = \dfrac{11,232,000}{17,576,000}$
$\qquad\qquad = (0.639)$ or 63.9%

Example 4: Suppose the result of an experiment has a sample space with 21 equally-likely outcomes, the consecutive integers from -10 to 10.

Let the event E = {-3,-2,-1,0,1,2,3,4,5} and
Let the event F = {-7,-6,-5,-4,-3,-2,-1}.
(E ∩ F) = {-3,-2,-1}.

What is the Pr(E|F) and the Pr(F|E)?

Pr(E|F) = $\frac{Pr(E \cap F)}{Pr(F)}$

$= \left[\frac{card(E \cap F) / card(sample\ space)}{card(F) / card(sample\ space)} \right] = \left(\frac{3/21}{7/21} \right) = 3/7.$

Pr(F|E) = $\frac{Pr(E \cap F)}{Pr(E)}$

$= \left[\frac{card(E \cap F) / card(sample\ space)}{card(E) / card(sample\ space)} \right] = \left(\frac{3/21}{9/21} \right) = 1/3.$

Example 5: Two cards are drawn without replacement from a well-shuffled 52 card deck. Let E be the event that the first card drawn is a "6 of clubs," and let F be the event that the second card drawn is a "jack" or a "6." What is the Pr(E ∩ F)?

From the general multiplication rule:
The Pr(E ∩ F) = Pr(E) · Pr(F|E) = $\left(\frac{1}{52} \right) \cdot \left(\frac{7}{51} \right) = \frac{7}{2652} \approx 0.00264.$

Example 6: Three computers, call them A, B, and C, must be working at the same time for the entire system to operate properly. All three computers function independently of each other. Computer A is operating at random 92% of the time. Computer B is operating at random 97% of the time. computer C is operating at random 94% of the time. What is the probability that at any given time all three computers are operating?

Since the computers are operating independently, the multiplication rule for independent events says that all three will be operating together $(.92)(.97)(.94) \cdot 100\%$ of the time, or about 83.9% of the time.

Example 7: If the result of an experiment is one of 15 equally-likely outcomes, and we take a simple random sample of size n = 8, how many simple random samples are there and what is the probability of each?

The experiment can result in one of 15 different outcomes in each repetition, and each with probability $\frac{1}{15}$ of occurrence. Since each repetition is independent of the others, each outcome within a sample can be considered to be an independent event. So the FPC says that there are $(15)^8 = 2,562,890,625$ different simple random samples, and the multiplication rule for independent events says that each sample has a probability of occurrence of $(\frac{1}{15})^8 \approx .00000000039$.

Example 8: If we have a well-shuffled 52 card deck and we deal a 5-card hand in the usual way, whenever we get 2 cards of the same denomination and 3 cards of a different denomination (such as 2 jacks and 3 aces), we call this a full house. What is the probability of being dealt a full house?

We first consider the number of ways that we can have only 2 denominations. There are C(13,1) possibilities for the first denomination. Then there are C(12,1) possibilities for the second denomination. Let's say that the two cards that are alike are from the first denomination and the other 3 cards that are alike are from the second denomination. There are C(4,2) possibilities for the two cards in the first denomination and C(4,3) possibilities for the three cards in the second denomination.

Thus there are $C(13,1) \cdot C(12,1) \cdot C(4,2) \cdot C(4,3)$
$= (13)(12)(6)(4) = 3744$ possible full houses.

There are C(52,5) = 2,598,960 equally-likely possible hands in the sample space. Calling the event of a full house E, the

$$Pr(E) = Pr(\text{full house}) = \frac{card(E)}{card(sample\ space)} = \frac{3744}{2,598,960} \approx 0.00144.$$

Example 9: An application of Bayes Theorem. Suppose that in a certain part of a city it is believed that any chosen person either has the HIV virus (event A_1) with probability (0.04), or doesn't have the HIV virus (event A_2), with probability (0.96). Obviously A_1 and A_2 are mutually exclusive and exhaustive events.

Some medical doctors in this city don't agree with these probabilities, so they intend to use a new test to find some updated probabilities. In other words, the doctors want to update the prior probabilities given above. This new test is believed to be much better than the old test. The new test gives a false positive result with the low probability (0.02), and a false negative with the probability (0.05). The event E is a positive test result, and the event $\widehat{E}$ is a negative test result. Therefore the $Pr(E|A_2) = 0.02$, and $Pr(E|A_1)$ = 0.98, and the $Pr(\widehat{E}|A_1)$ = 0.05, and $Pr(\widehat{E}|A_2) = 0.95$.

So a person is randomly chosen from this section of the city and tested. The person is found to be negative, as would most likely be the case. We wish to know the posterior probabilities $\{Pr(A_1|\widehat{E}), Pr(A_2|\widehat{E})\}$. Note we condition on $\widehat{E}$ since that was the result of the experiment.

Firstly, the $Pr(\widehat{E}) = Pr(\widehat{E}|A_1) Pr(A_1) + Pr(\widehat{E}|A_2)Pr(A_2)$
$$= (0.05)(0.04) + (0.95)(0.96)$$
$$= 0.002 + 0.912 = 0.914$$

So $\Pr(A_1 | \widehat{E}) = \frac{Pr(\widehat{E}|A_1)Pr(A_1)}{Pr(\widehat{E})} = \frac{0.002}{0.914} = 0.0022$

$\Pr(A_2 | \widehat{E}) = \frac{Pr(\widehat{E}|A_2)Pr(A_2)}{Pr(\widehat{E})} = \frac{0.912}{0.914} = 0.9978$

These posterior probabilities seem to indicate that the probability of a person in this section of the city having HIV is much lower than thought. However, this result is only from a sample of size one. The doctors will do a follow-up study with 50 people randomly chosen from this certain section of the city to get more accurate posterior probabilities.

CHAPTER 7:
PROBABILITY DISTRIBUTIONS

(41) **DISTRIBUTIONS**

A distribution is defined by a probability space. We know all of the outcomes from knowledge of the sample space $\mathbb{S}$. We can calculate the probability of any subset X of $\mathbb{S}$ from the probabilities of sets in the event space $\mathbb{E}$. The probability measure will give us the probability for any set in $\mathbb{E}$. For discrete distributions we have a formula, called a probability mass function (pmf), which assigns a non-negative number (a probability) to every outcome in $\mathbb{S}$, and therefore for every event in the event space $\mathbb{E}$. For continuous distribution models we have a probability density function (pdf), a curve in the coordinate plane, that allows us to calculate the probability of any event by way of a definite integral. The probability of a set E in $\mathbb{E}$ is the area under the pdf over the set E. For example if the event E = (a,b), the $\Pr(E) = \int_a^b f(x)dx$, which is the area under the pdf f(x) over the x-axis and between the x-values a and b.

(42) **EXPECTATION AND DISTRIBUTION PARAMETERS**

In my treatment of expected value in this section, and as a matter of my philosophy about sample spaces, we will assume that the sample space for any real-world distribution in actuality consists of a finite set of numbers $\{x_1, x_2, \ldots, x_N\}$. Therefore, our definitions of the expected values used to calculate the three main parameters of a distribution, the mean μ , the variance σ^2, and the standard deviation σ, will involve only the idea of a finite sample space and will involve only finite summations.

For those distribution models that we will encounter and use in the next chapters, which have countably infinite or uncountably infinite sample spaces, we will follow the conventional treatment. The population parameters for those distributions depend on the assumption of an infinite sample space, and that is what makes them possible in the first place and it is what makes them useful. According to the usual statistical theory, infinite sums and calculus are used to calculate them.

However, in this section, based on my philosophy, we will take the position that distributions based on infinite sample spaces are only models for underlying distributions that in actuality have only a finite sample space. Without doubt, probability models with infinite sample spaces are very convenient and very useful because they simplify the mathematics and allow us to easily calculate accurate probabilities. Nevertheless, here we will use only finite summations, and not use any infinite summations or calculus in the discussion of expected value and population parameters that follows below.

When we are dealing with a single distribution, we will of course be using only one summation in our definitions and use the notation $\Pr(X = x_i)$ to denote the probability that the distribution X takes the value x_i. When we are considering two distributions, we will encounter and use the notation $Pr(X = x_i, Y = y_j)$. This is called the joint pmf for distributions X and Y. It gives us the probability that X takes the value x_i and at the same time Y takes the value y_j. Note that if X and Y are distributions with sample spaces containing M and N outcomes respectively, then $\sum_{j=1}^{N} Pr(X = x_i, Y = y_j)$ holds X constant at the outcome x_i, while we sum over the probabilities for the outcomes in the sample space of Y, which must add up to one. So $\sum_{j=1}^{N} Pr(X = x_i, Y = y_j)$ equals the $Pr(X = x_i)$. Likewise, $\sum\limits_{i=1}^{M} Pr(X = x_i, Y = y_j) = Pr(Y = y_j)$.

Expectation, Distribution Parameters, and some Properties of Expectations:

(1) If we have a distribution X and the cardinality of the sample space is N, the expected value of X is defined to be:

$$E(X) = \sum\limits_{i=1}^{N} (x_i)Pr(X = x_i).$$

This turns out to be the mean value μ of the distribution. It's a weighted average of the numbers in $\mathbb{S}$.

We will also have use for the $E(X^2) = \sum\limits_{i=1}^{N} (x_i^2)\, Pr(X = x_i)$ in order to define the measures of dispersion known as the variance σ^2 and the standard deviation σ. This is the weighted average of the squares of the numbers in $\mathbb{S}$.

(2) For a real number a and for any distribution X, the E(a)

$$= \sum\limits_{i=1}^{N} (a)\, Pr(X = x_i) = (a) \sum\limits_{i=1}^{N} Pr(X = x_i) = (a)(1) = (a).$$

(3) If X is a distribution, and a $\in \mathbb{R}$, The $E(aX) = \sum\limits_{i=1}^{N} (a)(x_i)\, Pr(X = x_i)$

$$= (a) \sum\limits_{i=1}^{N} (x_i)\, Pr(X = x_i) = aE(X).$$

(4) If X and Y are distributions, and a,b $\in \mathbb{R}$, and the cardinality of the sample spaces for distributions X and Y is M and N, respectively,

then the $E(aX + bY) = \sum\limits_{i=1}^{M} \sum\limits_{j=1}^{N} (ax_i + by_j) \Pr(X = x_i, Y = y_j)$

$= \sum\limits_{i=1}^{M} \sum\limits_{j=1}^{N} (a)(x_i) \Pr(X = x_i, Y = y_j) + \sum\limits_{i=1}^{M} \sum\limits_{j=1}^{N} (b)(y_j) \Pr(X = x_i, Y = y_j)$

$= (a) \sum\limits_{i=1}^{M} (x_i) \sum\limits_{j=1}^{N} \Pr(X = x_i, Y = y_j) + (b) \sum\limits_{j=1}^{N} (y_j) \sum\limits_{i=1}^{M} \Pr(X = x_i, Y = y_j)$

$= (a) \sum\limits_{i=1}^{M} (x_i) \Pr(X = x_i) + (b) \sum\limits_{j=1}^{N} (y_j) \Pr(Y = y_j) = aE(X) + bE(Y).$

(5) If X and Y are independent distributions, and the cardinality of the sample spaces for X and Y are M and N, respectively, then the

$E(g(X)h(Y)) = \sum\limits_{i=1}^{M} \sum\limits_{j=1}^{N} g(x_i)h(y_j) \Pr(X = x_i, Y = y_j)$

$= \sum\limits_{i=1}^{M} \sum\limits_{j=1}^{N} g(x_i)h(y_j) \Pr(X = x_i) \cdot \Pr(Y = y_j)$ (independence of events)

$= \sum\limits_{i=1}^{M} g(x_i) \Pr(X = x_i) \cdot \sum\limits_{j=1}^{N} h(y_j) \Pr(Y = y_j) = E(g(X)) \cdot E(h(Y))$

Note that as a special case $E(XY) = E(X) \cdot E(Y)$, if X and Y are independent distributions.

(6) If $\{X_1, X_2, \ldots, X_n\}$ are independent distributions, and if

$X = (\sum\limits_{i=1}^{n} X_i)$ and $\overline{X} = (\frac{1}{n})(\sum\limits_{i=1}^{n} X_i)$, then the

$E(X) = E(\sum\limits_{i=1}^{n} X_i) = \sum\limits_{i=1}^{n} E(X_i) = n\mu$, and the

$E(\overline{X}) = E[(\frac{1}{n})(\sum\limits_{i=1}^{n} X_i)] = (\frac{1}{n})E(\sum\limits_{i=1}^{n} X_i) = (\frac{1}{n})(n\mu) = \mu.$

(7) For a distribution X, the variance(X) = σ^2 = Var(X) = $E[(X - \mu)^2]$
$= E(X^2) - 2\mu E(X) + E(\mu^2) = [E(X^2) - \mu^2] = E(X^2) - (E(X))^2.$

(8) For a real number a, the Var(a) = $E[(a - E(a))^2] = E[(a - a)^2] = 0$.

(9) If a $\in \mathbb{R}$, the Var(aX) = $E(a^2X^2) - (E(aX))^2 = [a^2E(X^2) - a^2(E(X))^2]$
$= (a^2)$ Var(X).

(10) If X and Y are independent distributions, and a,b $\in \mathbb{R}$, then the
Var(aX + bY) = $E[(aX + bY)^2] - [E(aX + bY)]^2$
$= E[a^2X^2 + 2ab(XY) + b^2Y^2] - (a\mu_x + b\mu_y)^2$
$= [a^2E(X^2) + 2abE(X)E(Y) + b^2E(Y^2)]$
$\qquad - a^2(\mu_x)^2 - 2ab(\mu_x)(\mu_y) - b^2(\mu_y)^2$
$= (a^2E(X^2) - a^2(\mu_x)^2) + (b^2E(Y^2) - b^2(\mu_y)^2)$
$= a^2$Var(X) + b^2Var(Y).

(11) If $\{X_1, X_2, \ldots, X_n\}$ are independent distributions, and if
$X = (\sum\limits_{i=1}^{n} X_i)$ and $\overline{X} = (\frac{1}{n})(\sum\limits_{i=1}^{n} X_i)$,

then the Var(X) = Var($\sum\limits_{i=1}^{n} X_i$) = $\sum\limits_{i=1}^{n}$ Var(X_i) = $n\sigma^2$, and

the Var($\overline{X}$) = Var[$(\frac{1}{n})(\sum\limits_{i=1}^{n} X_i)$] = $(\frac{1}{n^2})$Var($\sum\limits_{i=1}^{n} X_i$) = $(\frac{1}{n^2})(\sum\limits_{i=1}^{n}$ Var(X_i))

$= (\frac{1}{n^2})(n\sigma^2) = \frac{\sigma^2}{n}$.

(12) The Standard Deviation of a distribution X is
SD(X) = $\sigma = \sqrt{(Var(X))}$.

(13) For two distributions X and Y with means (μ_x) and (μ_y)
respectively, the covariance of X and Y, is defined to be
Cov(X,Y) = $E[(X - (\mu_x))(Y - (\mu_y))]$. This is a measure of how
X and Y vary together.

A positive value for Cov(X,Y) means that X and Y tend to increase together or decrease together, so that $(X - \mu_x)$ and $(Y - \mu_Y)$ tend to be positive together or negative together.

A negative value for Cov(X,Y) means that as X increases, Y tends to decrease, or vice versa. So that a positive value for $(X - \mu_x)$ tends to be associated with a negative value for $(Y - \mu_Y)$, and vice versa.

(14) The correlation of X and Y, $(\rho_{XY}) = \frac{Cov(X,Y)}{SD(X) \cdot SD(Y)}$, is a measure of the strength of the linear relationship between distributions X and Y. It turns out that $-1 \leq (\rho_{XY}) \leq 1$. When an experiment measures two variables X and Y, we have a set of data $\{(X_i, Y_i), i = 1, \ldots, n\}$. The closer that the pairs of (X_i, Y_i) data fit a straight line with a negative slope, the closer that (ρ_{XY}) is to -1. The closer that the pairs of (X_i, Y_i) data fit a straight line with a positive slope, the closer that (ρ_{XY}) is to +1. Note that (ρ_{XY}) is not the slope of a best fitting line, but a measure of how closely the (X_i, Y_i) data points fit the line.

(43) <u>MOMENT GENERATING FUNCTIONS (MGF)</u>

There is a very special type of expectation that we will consider for a distribution X. It is called the moment generating function of X. Note that the mgf for a distribution X, $M_X(t)$, is a function of a real variable t. Note that we can write M(t) for $M_X(t)$ when it is clear what the name of the distribution is.

(1) $M_X(t) = E(e^{tX}) = \sum_{i=1}^{N} e^{tX} \Pr(X = x_i)$. This is the first formulation.

Remember from calculus that $e^x = \sum\limits_{n=0}^{\infty} \frac{x^n}{n!}$. So then,

(2) $M_X(t) = E(e^{tX}) = E(\sum\limits_{n=0}^{\infty} (\frac{(tX)^n}{n!}))$

$= E(1 + (\frac{tX}{1!}) + (\frac{(t^2)(X^2)}{2!}) + (\frac{(t^3)(X^3)}{3!}) + \cdots)$

$= (1 + (\frac{tE(X)}{1!}) + (\frac{(t^2)(E(X^2))}{2!}) + (\frac{(t^3)(E(X^3))}{3!}) + \cdots)$.

This is the second formulation.

If we repeatedly take derivatives of the second formulation with respect to t, we find that the nth derivative of $M_X(t)$, evaluated at t = 0, is $E(X^n)$. These expectations, for n = 1,2,3, . . ., are called the moments of the distribution. In particular, $M'(0) = E(X)$, and $M''(0) = E(X^2)$. So if we can find the first formulation, we can take advantage of the properties found from the second formulation, which will allow us to easily find the mean and variance of the distribution from $E(X)$ and $E(X^2)$. An important property of an mgf is that if it exists in an open interval containing 0, then it is unique for that distribution. So mgf's can help us calculate distribution parameters and it can be used to identify a distribution. We will do an analysis with an mgf in a later chapter to prove a very important result in statistics known as the Central Limit Theorem (CLT). When proving the CLT, we use an mgf to identify a distribution.

An important result for mgf's is:
If $\{X_1, X_2, . . . ,X_n\}$ are independent and identically distributed, each with mgf $M^*(t)$, and $X = \sum\limits_{i=1}^{n} (X_i)$, then the mgf of X is:

$M_X(t) = E(e^{tX}) = E\left(e^{(t\sum\limits_{i=1}^{n}(X_i))}\right)$

$= E[(e^{(t(X_1))}) \cdot (e^{(t(X_2))}) \cdots (e^{(t(X_n))})]$

$= E(e^{(t(X_1))}) \cdot E(e^{(t(X_2))}) \cdots E(e^{(t(X_n))})$

$$= \prod_{i=1}^{n} M^*(t) = (M^*(t))^n .$$

This result is important because in probability and statistics we often deal with sums of independent distributions. So we use it to find the mgf of the sum as the product of the individual mgf's.

CHAPTER 8:
COMMON PROBABILITY
DISTRIBUTIONS

(44) <u>THE BERNOULLI DISTRIBUTION</u>

The first distribution that we will consider is the Bernoulli Distribution. If $\mathbb{S} = \{0, 1\}$, $\mathbb{E}$ is all possible subsets of $\mathbb{S}$, and the probability measure is $Pr(x) = \{ (p)$ if $x = 1$, $(1 - p)$ if $x = 0 \}$, where $0 \le p \le 1$, then we call this a Bernoulli(p) probability space, and the distribution it defines a Bernoulli(p) distribution. Sometimes we refer to the outcomes 0 and 1 generically as failure and success, respectively. Note that there are many different bernoulli(p) distributions for different choices of p. Often it is useful to model one repetition of an experiment with this distribution.

The mgf for X is $M(t) = E(e^{tX}) = \sum\limits_{x=0}^{1} (e^{tx})Pr(X = x)$

$= (e^{t(0)})(1 - p) + (e^{t(1)})(p)$. So, $M(t) = [(1 - p) +(p)e^{t}]$.

This mgf is obviously defined for all real numbers, so therefore it will be defined in an open interval containing 0, which is what we require.

$M'(t) = M''(t) = pe^{t}$.
$M'(0) = (p)e^{(0)} = p$. $M''(0) = (p)e^{(0)} = p$.

This says that $E(X) = E(X^2) = p$, for the bernoulli(p) distribution.
So the mean $\mu = p$, and the variance $\sigma^2 = E(X^2) - [E(X)]^2$
$= (p - p^2) = p(1 - p)$.

(45) <u>THE BINOMIAL DISTRIBUTION</u>

If $\mathbb{S} = \{0, 1, 2, . . . ,n\}$, and $\mathbb{E}$ is all the possible subsets of $\mathbb{S}$, and the $Pr(x) = \{ C(n,x)(p)^{x}(1 - p)^{n-x}$, for $x = 0,1,2,. . .,n\}$, where $0 \le p \le 1$, then we call this a Binomial(n,p) probability space, and the distribution it defines a binomial(n,p) distribution. Note that there is a different binomial distribution for different choices of n and p. If $\{X_1, X_2, ..., X_n\}$ are independent bernoulli distributions, then if $X = \sum\limits_{i=1}^{n} (X_i)$, then X has a binomial(n,p) distribution. X is the sum of n independent bernoulli(p) distributions. This X is the first of two statistics of interest here. Note X = the number of successes in the n trials. This is a whole number from 0 to n. The second statistic of interest here is $\overline{X} = (\frac{1}{n})(\sum\limits_{i=1}^{n} (X_i))$, which is the proportion of successes in the n trials. In this situation, $\overline{X}$ takes the outcomes $\{0, \frac{1}{n}, \frac{2}{n}, ..., 1\}$.

The moment generating function for X is

$M(t) = E(e^{tX}) = \sum\limits_{x=0}^{n} (e^{tx}) \cdot C(n,x)(p)^x(1-p)^{n-x} = \sum\limits_{x=0}^{n} C(n,x)(pe^t)^x(1-p)^{n-x}$

$= (pe^t + (1-p))^n$.

Note that since X is a sum of n independent bernoulli(p) distributions, then the mgf of X is the product of n bernoulli(p) mgf's. This is an application of the last result that we proved in the section above on mgf's. Since $0 \leq p \leq 1$, e^t is always positive, and n is a positive whole number, it is clear that this mgf will be defined for all real numbers, so therefore it will be defined in an open interval containing 0, which is what we require.

$M'(t) = (n)(pe^t + (1-p))^{n-1}(pe^t)$
$M''(t) = (np)[(pe^t + (1-p))^{n-1}(e^t) + (e^t)(n-1)(pe^t + (1-p))^{n-2}(pe^t)]$

So $E(X) = M'(0) = np$
$E(X^2) = M''(0) = n^2p^2 - np^2 + np$
So the mean $\mu = np$, and
the variance $\sigma^2 = (n^2p^2 - np^2 + np) - (np)^2 = n(p)(1-p)$.

Example 1: A woman has 4 pregnancies, and we assume only one child per pregnancy and no miscarriages. Let X = the number of successes = the number of boys. This can be (0,1,2,3, or 4). We'll summarize the probability distribution of X in a table.

x	Pr(X = x)
0	$C(4,0)(\frac{1}{2})^0(\frac{1}{2})^4 = (\frac{1}{16})$
1	$C(4,1)(\frac{1}{2})^1(\frac{1}{2})^3 = (\frac{1}{4})$
2	$C(4,2)(\frac{1}{2})^2(\frac{1}{2})^2 = (\frac{3}{8})$
3	$C(4,3)(\frac{1}{2})^3(\frac{1}{2})^1 = (\frac{1}{4})$
4	$C(4,4)(\frac{1}{2})^4(\frac{1}{2})^0 = (\frac{1}{16})$

This turns out to be a symmetrical distribution because the Pr(success) = Pr(Failure) = $(\frac{1}{2})$. The probability of a specific outcome of (x) boys and (4-x) girls is $(\frac{1}{2})^x(\frac{1}{2})^{4-x} = (\frac{1}{2})^4 = (\frac{1}{16})$, for x = 0, 1,2,3,4. The factor C(4,x) tells us the number of ways that (x) boys and (4-x) girls can occur. For example, the outcomes where 1 boy and 3 girls occurs is represented by the members of the set

{BGGG, GBGG, GGBG, GGGB}, which has C(4,1) = 4 members.

The mean of a binomial(n = 4, p = $\frac{1}{2}$) distribution is np = (4)($\frac{1}{2}$) = 2. The variance of a binomial(n = 4, p = $\frac{1}{2}$) distribution is np(1-p) = (4)($\frac{1}{2}$)($\frac{1}{2}$) = 1. Therefore, the standard deviation = 1 also.

Example 2: On an automated assembly line that produces 100 motors per day, n = 10 are randomly selected for inspection at the end of the day and labeled as defective or non-defective. It is believed that the probability of a defective motor is ($\frac{1}{50}$). If the quality control technician finds 2 or more defective motors in the sample of 10, the entire production of 100 motors produced that day must be inspected. Note here that a success means defective. We wish to know the probability that 2 or more defectives will be found in the sample. The number of defectives X is a binomial(10, $\frac{1}{50}$) distribution, and has outcomes {0,1,2,3,4,5,6,7,8,9,10}. The Pr(2 or more defective motors) = 1 - Pr(0 or 1 defectives). This is true because the outcomes are disjoint events and the Pr(0) + Pr(1) + Pr(2) + $\cdots$ + Pr(10) = 1.

The Pr(0) = C(10,0)($\frac{1}{50}$)^0($\frac{49}{50}$)^{10} = 0.8171.
The Pr(1) = C(10,1)($\frac{1}{50}$)^1($\frac{49}{50}$)^9 = 0.1667.

So the probability of 2 or more defectives is
(Pr(2) + Pr(3) + Pr(4) + Pr(5) + Pr(6) + Pr(7) + Pr(8) + Pr(9) + Pr(10)) = 1 - (Pr(0) + Pr(1)) = 1 - 0.8171 - 0.1667 = 0.0162. So about 1.62% of the time the company will inspect all 100 of the days production of motors to try to ascertain if there are any problems with their equipment.

Example 3: A NASCAR driver has a $\frac{1}{10}$ chance of winning any one of the 20 races that he intends to compete in this year. We can assume that his performances in all 20 races are independent of each other. What is the probability that he wins exactly 1 or 2 of the 20 races?

We can model this with a binomial(n = 20, p = $\frac{1}{10}$) distribution.
The Pr(1 win) = C(20,1) $\cdot$ ($\frac{1}{10}$)^1($\frac{9}{10}$)^{19} = (20)(.1)(.9)^{19} = 0.27
The Pr(2 wins) = C(20,2) $\cdot$ ($\frac{1}{10}$)^2($\frac{9}{10}$)^{18} = (190)(.1)^2(.9)^{18} = 0.2852
Therefore the Pr(He wins exactly 1 or 2 of the 20 races)
= (0.27 + 0.2852) = 0.5552 = 55.52%.

(46) THE GEOMETRIC DISTRIBUTION

If $S = \{1, 2, 3, \ldots\}$, and E is all possible subsets of S, and $Pr(x) = (1 - p)^{x-1}(p)$ for the outcomes in S, then this is called a Geometric(p) probability space, and the distribution it defines a geometric(p) distribution. There is a different geometric(p) distribution for all possible choices of p. This distribution gives us the probability that the first success occurs on the xth trial.

The mgf for this distribution is $M(t) = E(e^{tX}) = \sum\limits_{x=1}^{\infty} (e^{tx})(1 - p)^{x-1}(p)$

$= \left(\frac{p}{1-p}\right) \sum\limits_{x=1}^{\infty} ((1 - p)(e^t))^x$. If $|(1 - p)(e^t)| < 1$, meaning $t < \ln\left(\frac{1}{1-p}\right)$, then

this mgf exists on the interval $(-\infty, \ln\left(\frac{1}{1-p}\right))$, which contains 0, and this is what we require. The above series for t in this interval, is a convergent geometric series, and it has the sum $[(\frac{1}{1-(e^t-pe^t)}) - 1)]$.

So $M(t) = \left(\frac{p}{1-p}[(\frac{1}{1-(e^t-pe^t)}) - 1)]\right) = \left(\frac{pe^t}{(1-e^t+pe^t)}\right)$.

M'(t) can be calculated to be $\left(\frac{pe^t}{(1-e^t+pe^t)^2}\right)$.

So $M'(0) = E(X) = \left(\frac{1}{p}\right)$ is the mean.

M''(t) can be calculated to be $\left(\frac{(pe^t)(1+e^t-pe^t)}{(1-e^t+pe^t)^3}\right)$.

So $M''(0) = \left(\frac{2-p}{p^2}\right)$.

Therefore, the $Var(X) = M''(0) - (M'(0))^2 = (E(X^2) - \mu^2) = \frac{(1-p)}{p^2}$.

Example 1: Suppose we have a two outcome sequential process (the outcomes are generically called success and failure). The Pr(success)

$= p = 0.09$ and the Pr(failure) = (1 - p) = 0.91. What is the probability that the first success occurs on the seventh trial?

This probability is the Pr(6 failures and then a success) which equals $(1 - p)^6(p) = (.91)^6(.09) = 0.0511$.

Example 2: Biologists working at the city zoo are trying to mate two Panda bears. They know from past experience that the Pr(success) = Pr(the offspring will survive the first month of life) = (0.05). So the Pr(failure) = (0.95). Each pregnancy is assumed to be independent of the others. So we will model this situation with a geometric(0.05) distribution. We wish to know the Pr(a success occurs in the first five attempts). The geometric distribution probabilities say that the chance of a success in the first five attempts is

$$\sum_{x=1}^{5} (0.95)^{x-1}(0.05)$$

$= (.05) + (.95)(.05) + (.95)^2(.05) + (.95)^3(.05) + (.95)^4(.05)$
$= (.05) + (.0475) + (.045125) + (.04286875) + (.040725312)$
$= 0.2262$.

Example 3: An accountant has a probability of 0.35 of passing the CPA exam on any given try. This is the Pr(success). So the Pr(failure) = 0.65 on any single try. The attempts are independent of each other. What is the probability that the accountant will pass the test within the first three tries? The geometric distribution probabilities say that the

chance of success in the first three attempts is $\sum_{x=1}^{3} (0.65)^{x-1}(0.35)$

$= (0.35) + (0.65)(0.35) + (0.65)^2(0.35)$
$= (.35) + (.2275) + (.147875)$
$= 0.7254$.

(47) THE POISSON DISTRIBUTION

If $\mathbb{S} = \{0, 1, 2, \ldots\}$, and $\mathbb{E}$ is all subsets of $\mathbb{S}$, and $\Pr(x) = (e^{-\mu})(\frac{\mu^x}{x!})$ for the outcomes in $\mathbb{S}$, then this is called a Poisson(μ) probability space, and the distribution it defines a poisson(μ) distribution, where $\mu > 0$. There is a different poisson(μ) distribution for all the possible choices of μ. If we have a process where we are interested in the number of occurrences of some event during a particular interval of time, or in a particular region of space, etc., and the occurrences are random and at a constant average rate μ, then we say that we have a poisson process. The probability of x occurrences within the interval of time or in the region of space is given by the above probability measure.

If we assume the interval of time or region of space is divided equally into a large number n of small parts, where we assume that the probability of an occurrence p is the same for each small part, and that only one occurrence can happen within each small part, then the total number of occurrences can be modeled with a binomial distribution X. Under these assumptions, we can show that poisson probabilities can be used to approximate binomial probabilities when n goes to infinity and p goes to zero in such a way that the mean number of successes for X remains constant:

The binomial pmf is $p(x) = C(n,x)p^x(1-p)^{n-x}$

$$= \frac{n!}{(n-x)!(x!)}p^x(1-p)^{n-x} = \frac{n!}{(n-x)!(x!)}\frac{n^x}{n^x}p^x(1-p)^{n-x}$$

$$= \frac{n!}{(n-x)!} \cdot \frac{1}{(n-np)^x} \cdot \frac{(np)^x}{x!} \cdot \left(1 - \frac{(np)}{n}\right)^n$$

$$= \left(\frac{n}{n-np}\right)\left(\frac{n-1}{n-np}\right)\cdots\left(\frac{n-(x-1)}{n-np}\right) \cdot \left(1 - \frac{(np)}{n}\right)^n \cdot \frac{(np)^x}{x!} \; .$$

Now let n $\to \infty$ and p $\to 0$, in such a way that $\mu = (np)$ remains constant. The first x factors will tend to one, since (np) and (x) remain small, relative to n. So this converges to $p(x) = e^{-\mu} \cdot \frac{\mu^x}{x!}$. The mean $\mu = (np)$ remains relatively small and constant, so that the poisson probabilities can be used to approximate the binomial probabilities.

The mgf of this distribution is $M(t) = E(e^{tX}) = \sum\limits_{x=0}^{\infty} (e^{tx})(e^{-\mu})(\frac{\mu^x}{x!})$

$= (e^{-\mu})(e^{\mu e^t}) \sum\limits_{x=0}^{\infty} e^{-\mu e^t} \cdot (\frac{(\mu e^t)^x}{x!})$

$= (e^{\mu e^t - \mu}) \cdot \sum\limits_{x=0}^{\infty} e^{-\mu e^t} \cdot (\frac{(\mu e^t)^x}{x!})$. The summation is 1, because

it is the sum of the probabilities for a poisson(μe^t) distribution.

So $M(t) = (e^{\mu e^t - \mu})$, and it is defined for all t $\in \mathbb{R}$, which contains 0, which is what we require.

$M'(t) = (e^{\mu e^t - \mu}) \cdot (\mu e^t)$.
So $M'(0) = E(X) = \mu$ is the mean.

$M''(t) = (e^{\mu e^t - \mu}) \cdot (\mu e^t) + (e^{\mu e^t - \mu}) \cdot (\mu e^t)^2$
$\qquad = (e^{\mu e^t - \mu}) \cdot (\mu e^t) \cdot (1 + \mu e^t)$.
So $M''(0) = E(X^2) = \mu + \mu^2$.

Therefore, the Var(X) = $M''(0) - (M'(0))^2$ = $E(X^2) - (E(X))^2$
$\qquad\qquad\qquad = (\mu + \mu^2) - \mu^2 = \mu$.

Therefore, the standard deviation of X is $\sqrt{\mu}$.

Example 1: The number of car accidents per month at a particular intersection is modeled as a poisson distribution X with mean number of occurrences $\mu = 3$. What is the probability of 6 accidents in a month?

The $Pr(X = 6 \mid \mu = 3) = (e^{-3})(\frac{3^6}{6!}) = (0.049787068)(\frac{729}{720}) \approx 0.0504$.

Example 2: The number of Canadian Oak pollen grains falling on an experimental test area of one square millimeter per day, from a tree of this species 100 yards distant, is modeled as a poisson distribution with mean $\mu = 2$. The number of grains that fall on the test area on a given day is independent of the number that fall on other days. So we have a Poisson distribution X with mean $\mu = 2$. What is the $Pr(X \geq 2)$?

$$Pr(X \geq 2 \mid \mu = 2) = \sum_{n=2}^{\infty} Pr(X = n) = 1 - \sum_{n=0}^{1} Pr(X = n)$$

$$= 1 - Pr(X = 0) - Pr(X = 1) \quad (\text{since } \sum_{n=0}^{\infty} Pr(X = n) = 1)$$

$$= 1 - (e^{-2})(\frac{2^0}{0!}) - (e^{-2})(\frac{2^1}{1!}) \approx 0.594.$$

(48) THE NORMAL DISTRIBUTION

Here we will consider a distribution X with mean μ and variance σ^2, which has a large finite number N of outcomes in its sample space $\mathbb{S} = \{x_1, x_2, \ldots, x_N\}$, and $\mathbb{E}$ is the set of all possible subsets of $\mathbb{S}$, and the $Pr(X = x) = (\frac{1}{N})$ for each of the N outcomes. This naturally leads to the counting measure for probability, which we have seen before.

The outcomes, though equally likely, are not spread out evenly on the real line and are usually very large in number, so that we may not know what all the outcomes are. All that we may know is that the outcomes in the sample space tend to be very densely located near the mean of the distribution, and the density falls off symmetrically and rapidly as we move away from the mean on either side. The outcomes are mound shaped near a central value. This is the way that many real world data sets are actually distributed. So what statisticians do is assume that the density of the outcomes of the distribution X is approximated closely by a so-called probability density function (pdf) known as the normal curve, and we then say that X is normally distributed with mean μ and standard deviation σ. This is a continuous model for the underlying distribution X. The normal curve f(x), with these parameters μ and σ, is

$$f(x) = \left(\frac{1}{\sqrt{2\pi}\,\sigma} \cdot e^{-\frac{(x-\mu)^2}{2\sigma^2}} \right) , \quad -\infty < x < \infty.$$

This function is also commonly known as the bell-shaped curve, and is used to model the distribution of many sets of data in applied statistics. Figure 13 shows a normal curve where the mean $\mu = 10$ and the standard deviation $\sigma = 2$.

Figure 13

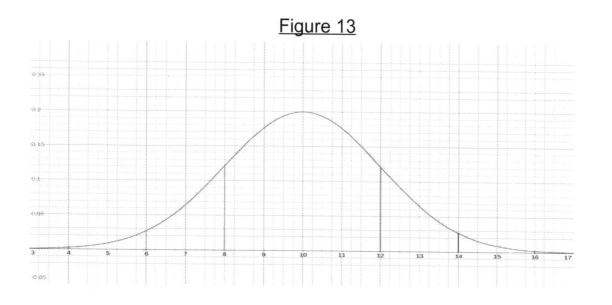

The $\Pr(\mu - \sigma < X < \mu + \sigma) = \Pr(8 < X < 12) = 68.3\%$
The $\Pr(\mu - 2\sigma < X < \mu + 2\sigma) = \Pr(6 < X < 14) = 95.4\%$

The Pr($\mu - 3\sigma < X < \mu + 3\sigma$) = Pr(4 < X < 16) = 99.7%

The sample space for the normal distribution is $\mathbb{S} = (-\infty, \infty)$ and the event space $\mathbb{E}$ is all positively measured subsets of $\mathbb{S}$ (intervals). We are not able to use the counting measure, so instead we get a measure of the set (a,b) by integrating f(x) between a and b:

$$Pr(a < X < b) = \int_a^b f(x)dx = \int_a^b \left(\frac{1}{\sqrt{2\pi}\,\sigma} \cdot e^{-\frac{(x-\mu)^2}{2\sigma^2}}\right)\ dx\ .$$

This often turns out to be an excellent and very useful approximation which we can use to get probabilities for many distributions. This particular function f(x) does not have an antiderivative in closed form. Integrating it would require a numerical technique on a computer. Actually, when we want to find the probability of a real set (usually an interval), we make use of a standardized distribution that we call the Z distribution.

We define $Z = \left(\frac{X-\mu}{\sigma}\right)$.

If X is normally distributed, then the distribution of Z is also normal, but with mean $\mu = 0$ and $\sigma = 1$, which can be shown using the formulas for the mean and variance of a distribution in chapter 7.

Z follows a normal curve g(z) found by substituting $\mu = 0$ and $\sigma = 1$ in the above function f(x) and changing the independent variable to z. The distribution of Z is called the standard normal distribution (Z distribution) and has pdf:

$$g(z) = \left(\frac{1}{\sqrt{2\pi}} \cdot e^{-\frac{(z)^2}{2}}\right),\quad -\infty < z < \infty.$$

We use a computer to approximate the function $G(z) = \int\limits_{-\infty}^{z} g(t)\, dt$,

for z-values from about -4 to 4. These G(z)-values are then put into table form, and we call this the Table of the Standard Normal Distribution. The Pr(Z < -4) ≈ 0.0000 and the Pr (Z < 4) ≈ 1.0000. This says that the greater part of the Z distribution is between -4 and 4, because the greater part of any normal distribution X is within 4 standard deviations above and below the mean value, and that makes the table of the standard normal distribution very useful in applications. This table can be found in many applied statistics books and literature, and available for use with many statistical software packages. In this book, we will use probabilities from the Z distribution extensively, however, we will not include a Z distribution table in this book. The reader will have to take it on faith that all Z distribution probabilities used in this book were correctly taken from an external source.

Note that the G(z)-values are the area (probability) that has accumulated from (-∞) up to z. We call G(z) a cumulative distribution function (cdf). Figure 14 shows the standard normal distribution curve ($\mu = 0$, $\sigma = 1$).

Figure 14

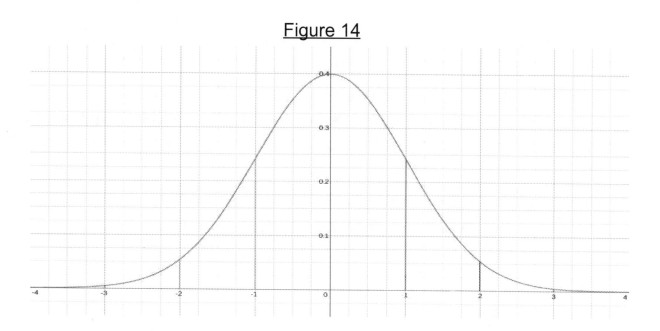

The Pr($\mu - \sigma < Z < \mu + \sigma$) = Pr(-1 < Z < 1) = 68.3%

The Pr($\mu - 2\sigma < Z < \mu + 2\sigma$) = Pr(-2 < Z < 2) = 95.4%

The Pr($\mu - 3\sigma < Z < \mu + 3\sigma$) = Pr(-3 < Z < 3) = 99.7%

We use the Z distribution in the following way:

Example 1: Suppose X has a normal distribution with mean μ = 100 and σ = 15. The probabilities for the standard normal distribution that we will use from here on out were obtained from an external source.

(A) To find the Pr(80 < X < 105), we note that this is the same as the

$\quad$ Pr($\frac{80-100}{15}$ < Z < $\frac{105-100}{15}$) = Pr(-1.33 < Z < 0.33) = G(0.33) - G(-1.33)

$\quad$ = (0.62930) - (0.09176) = 0.53754.

(B) To find the Pr(X < 115) = Pr(Z < $\frac{115-100}{15}$) = Pr(Z < 1.00)

$\quad$ = G(1.00) = 0.84134.

(C) To find the Pr(X > 120) = Pr(Z > $\frac{120-100}{15}$) = Pr(Z > 1.33)

$\quad$ = 1 - Pr(Z < 1.33) = 1 - G(1.33) = 1 - (0.90824) = 0.09176.

The implication of all this is that we can find probabilities associated with any normal distribution X using these cdf probabilities for a Z distribution. Though X (with arbitrary parameters μ and σ) and Z (with parameters 0 and 1) are different normal distributions with different shapes, there is a one-to-one correspondence between X and Z.

This one-to-one correspondence is a function H(x) from $\mathbb{R}$ to $\mathbb{R}$, H(x): x $\rightarrow$ ($\frac{x-\mu}{\sigma}$). For example, H maps (μ) to 0, H maps ($\mu - \sigma$) to -1, H maps ($\mu + 2\sigma$) to 2, and so one. Since this is a 1-1 and onto function, there is an inverse function $H^{-1}(z)$ from $\mathbb{R}$ to $\mathbb{R}$ given by $H^{-1}(z)$: z $\rightarrow \mu + z\sigma$. H takes an X-value and returns a unique Z-value.

H^{-1} takes a Z-value and returns a unique X-value.

The mgf of a Normal distribution X with mean μ and variance σ^2 is

$$M(t) = E(e^{tX}) = \int_{-\infty}^{\infty} (e^{tX}) \cdot \left(\frac{1}{\sqrt{2\pi}\,\sigma} \cdot e^{-\frac{(x-\mu)^2}{2\sigma^2}} \right) dx \;.$$ This is a rather

formidable integral and it requires a lengthy calculation to evaluate it. We will simply state here that the result is:

$$M(t) = e^{(\mu t + (\frac{t^2 \sigma^2}{2}))}\;.$$ This is defined for all real numbers, so therefore it is defined in an open interval containing zero, which we require.

$$M'(t) = \left(e^{(\mu t + (\frac{t^2 \sigma^2}{2}))} \right) \cdot (\mu + t\sigma^2).$$

$$M''(t) = \left(e^{(\mu t + (\frac{t^2 \sigma^2}{2}))} \right)(\sigma^2) + (\mu + t\sigma^2)^2 \left(e^{(\mu t + (\frac{t^2 \sigma^2}{2}))} \right).$$

$M'(0) = E(X) = \mu.$
$M''(0) = E(X^2) = \sigma^2 + \mu^2$

So the mean $= \mu$, and the variance is $(\sigma^2 + \mu^2) - \mu^2 = \sigma^2$.
Then the standard deviation $= \sigma$.

For the standard normal distribution, the mgf can be found by setting $\mu = 0$ and $\sigma = 1$ in the mgf for a general normal distribution, which yields M(t) = $e^{\frac{t^2}{2}}$. This mgf will be important later.

A linear combination of two normal distributions:
To illustrate the fact that a linear combination of two or more normal distributions is normal, we will use the simple example of the sum of two normal distributions X_1 and X_2. If X_1 and X_2 are independent normal distributions with means (μ_1) and (μ_2) respectively, and

standard deviations (σ_1) and (σ_2) respectively, then ($X_1 + X_2$) is normally distributed also. We can use mgf's to see this:

$$M(t) = E\left(e^{t(X_1 + X_2)}\right) = E\left(e^{tX_1} \cdot e^{tX_2}\right)$$
$$= E\left(e^{tX_1}\right) \cdot E\left(e^{tX_2}\right)$$
$$= e^{(\mu_1 t + (\frac{t^2\sigma_1^2}{2}))} \cdot e^{(\mu_2 t + (\frac{t^2\sigma_2^2}{2}))}$$
$$= e^{((\mu_1 + \mu_2)t + ((\sigma_1^2 + \sigma_2^2)t^2)/2}.$$

This is the mgf of a normal (($\mu_1 + \mu_2$), (($\sigma_1)^2 + (\sigma_2)^2$)) distribution. Therefore ($X_1 + X_2$) has this normal distribution. A generalization of this result is that any linear combination of independent normal distributions such as, for example W = 3X - 10Y, where X and Y are independent, will also be a normal distribution. To find the mean and variance of W, assuming that we know the mean and variance of X and Y, we can use the expectation formulas from a previous section in chapter 7.

Example 2: Suppose that the number of barrels of oil X that is pumped out of a specific well per day is normally distributed with mean $\mu = 9800$ and standard deviation $\sigma = 50$. What is the $Pr(X \leq 9760)$?

The $Pr(X \leq 9760) = Pr(Z \leq \frac{9760 - 9800}{50}) = Pr(Z \leq -0.80)$
$$= G(-0.80) = 0.21186.$$

Example 3: On an IQ test developed by Professor Sanders, genius is defined to be a score of 140 or higher. The test scores X of 5000 people showed that scores on this test are normally distributed with a mean of 100 and a standard deviation of 16. What is the probability of being rated a genius according to Professor Sanders?

The $Pr(X \geq 140) = Pr(Z \geq \frac{140 - 100}{16}) = Pr(Z \geq 2.50) = 1 - Pr(Z \leq 2.50) =$

1 - G(2.50) = 1 - 0.99379 = 0.00621. This is 0.621% of the population.

Example 4: An automated soda machine dispenses beverages at the county fair, the amount of which is normally distributed with mean μ = 16 ounces and standard deviation σ = 0.2 ounces. What is the probability that a customer will get 15.8 to 16.3 ounces of soda from this machine?

If X equals the amount dispensed at this machine in a given fill, we wish to know the Pr(15.8 $\leq$ X $\leq$ 16.3) = Pr($\frac{15.8-16}{0.2}$ $\leq$ Z $\leq$ $\frac{16.3-16}{0.2}$)

= Pr(-1.00 $\leq$ Z $\leq$ 1.50) = Pr(Z $\leq$ 1.50) - Pr(Z $\leq$ -1.00)

= G(1.50) - G(-1.00) = 0.93319 - 0.15866

= 0.77453 = 77.453%.

(49) THE GAMMA FUNCTION

Before we discuss the so-called Gamma distribution, we will first discuss the Gamma function and a couple results concerning it. The gamma function depending on the parameter α is

$$\Gamma(\alpha) = \int_0^\infty x^{\alpha-1} e^{-x} dx.$$

Let's use integration by parts to get a useful result: Let u = $x^{\alpha-1}$ and

dv = $e^{-x}dx$. Then du = $(\alpha-1)x^{\alpha-2} dx$ and v = $\int e^{-x} dx = -e^{-x}$. So

$\Gamma(\alpha) = [(x^{\alpha-1})(-e^{-x})]_0^\infty + (\alpha-1)\int_0^\infty x^{(\alpha-1)-1}e^{-x}dx$. That is,

$\Gamma(\alpha) = (\alpha-1)\Gamma(a-1)$.

It turns out that for an integer $\alpha \geq 1$, $\Gamma(\alpha) = (\alpha-1)!$

A result that will be important is that $\Gamma(\frac{1}{2}) = \int_0^\infty x^{-\frac{1}{2}}e^{-x}dx$. Let x = $\frac{y^2}{2}$.

Then $\Gamma(\frac{1}{2}) = \int_0^\infty \frac{\sqrt{2}}{y} e^{-\frac{y^2}{2}} y \, dy = 2\sqrt{\pi} \int_0^\infty \frac{1}{\sqrt{2\pi}} e^{-\frac{y^2}{2}} dy$. The integral is $\frac{1}{2}$, since it is half of the area under the standard normal pdf. Therefore $\Gamma(\frac{1}{2}) = \sqrt{\pi}$.

(50) <u>THE GAMMA DISTRIBUTION</u>

The pdf of a Gamma distribution with parameters $\alpha > 0$ and $\beta > 0$ is:

$f(x) = \left(\frac{x^{\alpha-1} e^{-x/\beta}}{\Gamma(\alpha) \beta^\alpha} \right)$, $0 \le x < \infty$.

Figure 15 shows the pdf for a Gamma($\alpha = 2$, $\beta = 2$) distribution.

<u>Figure 15</u>

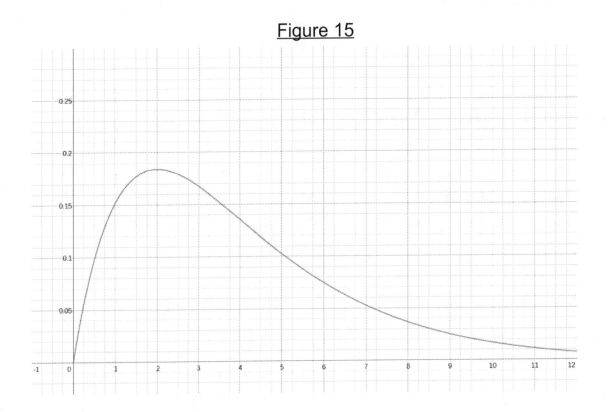

The mgf of a Gamma distribution X is:

$M(t) = E(e^{tX}) = \int_0^\infty (e^{tX}) \cdot \left(\frac{x^{\alpha-1} e^{-x/\beta}}{\Gamma(\alpha) \beta^\alpha} \right) \, dx$.

Once again as with the normal distribution, this is a lengthy computation. We will simply state that the result is:

$M(t) = (1 - \beta t)^{-\alpha}$.

This will be defined in an open interval containing zero if we make $1 - \beta t > 0$. This means $t < \frac{1}{\beta}$, where $\beta > 0$. So this mgf is defined in the interval $(-\infty, \frac{1}{\beta})$ which contains 0.

$M'(t) = (-\alpha)(1 - \beta t)^{-\alpha-1}(-\beta)$
$M''(t) = (\alpha\beta)((-\alpha - 1)(1 - \beta t)^{-\alpha-2}(-\beta)) = (\alpha\beta^2)((\alpha + 1)(1 - \beta t)^{-\alpha-2})$

$M'(0) = \alpha\beta$.
$M''(0) = \alpha^2\beta^2 + \alpha\beta^2$

So the mean of a gamma distribution is $\mu = \alpha\beta$, and
the variance of a gamma distribution is $\sigma^2 = M''(0) - (M'(0))^2$
$= (\alpha^2\beta^2 + \alpha\beta^2) - (\alpha^2\beta^2) = \alpha\beta^2$.

(51) THE CHI-SQUARE DISTRIBUTION

We will now show how the distribution of (Z^2) is a type of Gamma distribution.

If $X = (Z^2)$, and the cdf of $X = (Z^2)$ is F(x), then

$F(x) = Pr(Z^2 \le x) = Pr(-\sqrt{x} \le Z \le \sqrt{x}) = 2 \cdot Pr(0 \le Z \le \sqrt{x})$, because the Z distribution is symmetric about the y-axis (the line x = 0).

So $F(x) = 2 \int_{0}^{\sqrt{x}} (1/\sqrt{2\pi}) \cdot (e^{(-\frac{(t)^2}{2})})$ dt. We can find the pdf of X if we know that the pdf of X is f(x) = F'(x).

So $f(x) = (2) \cdot (\frac{1}{\sqrt{2\pi}})(e^{-\frac{x}{2}})(\frac{1}{2\sqrt{x}}) = \frac{e^{-x/2}}{\sqrt{2\pi}\sqrt{x}}$, for $0 \le x < \infty$.

(Note that from the chain rule, after we plug $\sqrt{x}$ in for t in the integrand, we have to multiply the result by $(\frac{d}{dx}(\sqrt{x})) = \frac{1}{2\sqrt{x}}$).

So $f(x) = (\frac{x^{1/2-1}e^{-x/2}}{\sqrt{\pi}\sqrt{2}})$, for $0 \le x < \infty$, is the pdf of a Gamma distribution with $\alpha = \frac{1}{2}$, $\beta = 2$, noting that $\Gamma(\alpha) = \Gamma(\frac{1}{2}) = \sqrt{\pi}$, and $(\beta)^{\alpha} = \sqrt{2}$.

We call a gamma ($\alpha = \frac{n}{2}$, $\beta = 2$) distribution a chi-square distribution with n degrees of freedom (df). So $X = (Z^2)$ has a chi-square distribution with one degree of freedom (df).

If we have the sum of n independent (Z^2) distributions, that is $X = \sum_{i=1}^{n} (Z_i)^2$, then let's find the mgf of X.

$M_X(t) = E(e^{tX}) = E\left(e^{t\sum_{i=1}^{n}(Z_i^2)} \right) = E((e^{t(Z_1^2)}) \cdot (e^{t(Z_2^2)}) \cdots (e^{t(Z_n^2)}))$

$= E(e^{t(Z_1^2)}) \cdot E(e^{t(Z_2^2)}) \cdots E(e^{t(Z_n^2)})$

$= \prod_{i=1}^{n} (M(t))$ (where M(t) is the mgf of (Z^2) distribution)

$= \prod_{i=1}^{n} (1-2t)^{-1/2} = (1-2t)^{-n/2}$.

Therefore, $X = \sum_{i=1}^{n} (Z_i)^2$ has a chi-square distribution with n df.

Figure 15 was the graph of a Gamma($\alpha = 2, \beta = 2$) distribution, which is the same as a Chi-square distribution with 4 df, since $\alpha = 2 = \frac{4}{2}$.

Normal and Chi-square distributions are very important in statistics. The mean of a Chi-square distribution with n df is $\alpha\beta = (\frac{n}{2})(2) = n$. The variance of a Chi-square distribution with n df is $\alpha\beta^2 = (\frac{n}{2})(2)^2 = 2n$.

(52) <u>THE EXPONENTIAL DISTRIBUTION</u>

There is another gamma distribution which has a lot of application. If we let $\alpha = 1$ and $\beta = \mu$, then we get the very simple gamma pdf:

$f(x) = \frac{1}{\mu}\, e^{-x/\mu}$, for x ≥ 0. We call this type of gamma distribution the Exponential distribution with parameter μ. The mean is $\alpha\beta = \mu$.
The variance is $\alpha\beta^2 = \mu^2$. So the standard deviation is μ.
Figure 16 is the graph of an exponential($\mu = 3$) distribution.

<u>Figure 16</u>

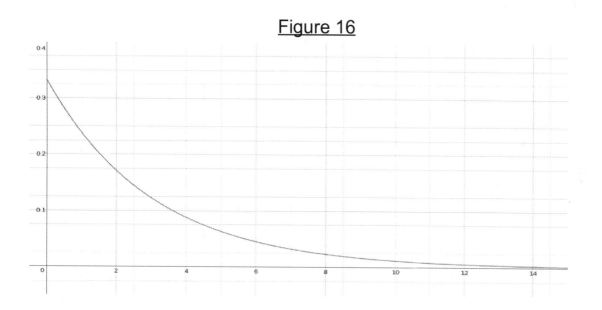

This distribution is important in many areas of applied probability, such as in queueing theory. If we have a poisson arrival process, then it turns out that inter-arrival times (times between arrivals) are exponentially distributed.

An important property of the exponential distribution is that it is so-called memoryless. This means that if X has an exponential distribution, the

$Pr(X > 2c \mid X > c) = Pr(X > c)$. An example in queueing theory is that if you have already waited a time c with no arrivals, then the probability that you have to wait another time c before the first arrival is the same as having to wait a time c for the first arrival from the beginning.

Mathematically, the $Pr(X > 2c \mid X > c) = \dfrac{Pr((X > 2c) \cap (X > c))}{Pr(X > c)} =$

$$\dfrac{\displaystyle\int_{2c}^{\infty} \tfrac{1}{\mu} e^{-x/\mu} dx}{\displaystyle\int_{c}^{\infty} \tfrac{1}{\mu} e^{-x/\mu} dx} = \dfrac{[e^{-x/\mu}]_{2c}^{\infty}}{[e^{-x/\mu}]_{c}^{\infty}} = \dfrac{0 - e^{-2c/\mu}}{0 - e^{-c/\mu}} = e^{-c/\mu} = Pr(X > c).$$

CHAPTER 9:
INTRODUCTION TO STATISTICS

(53) **SAMPLES**

Usually in scientific research we conduct an experiment where we collect a sample of n observations $\{x_1, x_2, \ldots, x_n\}$ from some distribution X in order to make statistical inferences. Statistical inference is often the making of a test of hypotheses or the construction of a confidence interval for a mean (μ) or a variance (σ^2). We will cover these topics in this chapter.

The sample values $\{x_1, x_2, \ldots, x_n\}$ are considered to be n independent observations from the distribution X. We say that we have a sample of size n. It is assumed that every sample of size n has the same chance of being selected as any other sample of size n. We say then that we are collecting a simple random sample (SRS).

(54) **SAMPLE STATISTICS**

A statistic is a function of the sample. Some statistics that we could calculate from a sample include the sample mean, median, mode, variance, standard deviation, range, inter-quartile range, and many more. The only statistics that we are going to concentrate on are the sample mean $\overline{X}$, the sample variance S^2, and the sample standard deviation S. The sample mean $\overline{X}$ estimates the distribution mean μ, the sample variance S^2 estimates the distribution variance σ^2, and the sample standard deviation S estimates the distribution standard deviation σ. These statistics vary from sample to sample, so they have a distribution. We call the distribution of a sample statistic its sampling distribution.

(1) The Distribution of the Sample Mean:

The sample average $\overline{X} = (\frac{1}{n})(\sum\limits_{i=1}^{n} X_i)$ is one of the most common and important statistics that we can compute from a sample.

(A) $E(\overline{X}) = (\frac{1}{n})E(\sum\limits_{i=1}^{n} X_i) = (\frac{1}{n})(\sum\limits_{i=1}^{n} E(X_i)) = (\frac{1}{n})(n\mu) = \mu$.

(B) $Var(\overline{X}) = (\frac{1}{n})^2 Var(\sum\limits_{i=1}^{n} X_i) = (\frac{1}{n^2})(\sum\limits_{i=1}^{n} Var(X_i)) = (\frac{1}{n^2})(n\sigma^2) = \frac{\sigma^2}{n}$.

Therefore as $n \to \infty$, $Var(\overline{X}) \to 0$.

Since $E(\overline{X}) = \mu$, then we say that $\overline{X}$ is an unbiased estimator of μ. Since $\lim\limits_{n\to\infty} Var(\overline{X}) = 0$, the $Pr(-\varepsilon < \overline{X} < \varepsilon) \to 1$ where ε is an arbitrarily small positive number. So $\overline{X}$ converges to the single point μ as $n \to \infty$.

(2) The Distribution of The Sample Variance:

The sample variance $S^2 = (\frac{1}{n-1})(\sum\limits_{i=1}^{n} (X_i - \overline{X})^2)$ is also one of the most common and important statistics that we can compute from a sample.

(A) $E(S^2) = E[(\frac{1}{n-1})(\sum\limits_{i=1}^{n} (X_i - \overline{X})^2)] = (\frac{1}{(n-1)}) \cdot E(\sum\limits_{i=1}^{n} (X_i - \overline{X})^2)$

$= (\frac{1}{n-1}) \cdot \sum\limits_{i=1}^{n} [(E(X_i)^2 - (\frac{2}{n})E(X_i(\sum\limits_{i=1}^{n} X_i)) + E(\overline{X})^2]$

$$= \left(\tfrac{1}{n-1}\right) \cdot \sum_{i=1}^{n} \left[(\mu^2 + \sigma^2) - \left(\tfrac{2}{n}\right)E(X_i)^2 - \left(\tfrac{2(n-1)}{n}\right)E(X_i)(X_j) + (\mu^2 + (\tfrac{\sigma^2}{n}))\right]$$

$$= \left(\tfrac{1}{n-1}\right) \cdot \left[(n\mu^2 + n\sigma^2) - (2)(\mu^2 + \sigma^2) - (2n-2)\mu^2 + (n\mu^2 + (\sigma^2))\right]$$

$$= \left(\tfrac{1}{n-1}\right) \cdot \left[(n-1)\sigma^2\right] = \sigma^2$$

(B) The statistic $\frac{(n-1)S^2}{\sigma^2} = \left(\tfrac{1}{\sigma^2}\right)\left(\sum_{i=1}^{n}(X_i - \overline{X})^2\right) = \sum_{i=1}^{n}\left(\frac{X_i - \overline{X}}{\sigma}\right)^2$.

As n increases, $\sum_{i=1}^{n}\left(\frac{X_i - \overline{X}}{\sigma}\right)^2$ has approximately a Chi-square distribution with n degrees of freedom because $\overline{X}$ converges to μ. Therefore the $\mathrm{Var}\left(\sum_{i=1}^{n}\left(\frac{X_i - \overline{X}}{\sigma}\right)^2\right) \approx 2n$.

So the $\mathrm{Var}(S^2) \approx \left(\frac{2n\sigma^4}{(n-1)^2}\right)$, which $\rightarrow 0$, as n increases.

The $E(S^2) = \sigma^2$, and the $\mathrm{Var}(S^2) \rightarrow 0$, as $n \rightarrow \infty$.

(3) The Distribution of the Sample Standard Deviation:

Another important statistic is the sample standard deviation S. $S = \sqrt{(Sample\ Variance)}$. It is clear that the $\lim_{n\to\infty} E(S) = \sigma$, and the $\lim_{n\to\infty} \mathrm{Var}(S) = 0$. Therefore, S is a good estimator of σ and is very useful in applied statistics.

(55) <u>THE CENTRAL LIMIT THEOREM (CLT)</u>

I have stated my position that all real-world distributions have finite sample spaces. Some discrete distributions like the geometric and the poisson have countably infinite sample spaces, but certainly not all of the outcomes could occur in any real world situation. The situation is similar with distributions that have uncountably infinite sample spaces. Any distribution with any kind of infinite sample space is simply a model for some underlying distribution with a finite sample space, but these models can be very useful nonetheless. In this section we consider distributions X with finite, but possibly very large, sample spaces, where the outcomes are the possible outcomes of some kind of experiment. Usually we assume that these distributions can be modeled well with the normal distribution curve, but this assumption is not necessary for the validity of this theorem. Since their sample spaces are actually finite, then the mean μ and the variance σ^2 and all moments about the mean $E(X-\mu)^k, \, for \, k \geq 3$, are finite. We will need to know this to prove the Central Limit Theorem as we do here.

<u>Statement and proof of the CLT:</u>
Suppose we have any distribution X with mean μ and variance σ^2. If we take a SRS from X and compute $\overline{X}$, then it turns out that the distribution of

$$Z = \left(\frac{\overline{X} - \mu}{\sigma/\sqrt{n}} \right) = \sqrt{n} \left(\frac{\overline{X} - \mu}{\sigma} \right)$$

converges to a standard normal distribution as n $\rightarrow \infty$. This is a very useful and important result known as the Central Limit Theorem (CLT). This theorem is valid for all distributions X as the sample size n gets large. As a rule of thumb, statisticians consider a sample size $n \geq 30$ to be a large sample. If we are sampling from approximately normal distributions then the CLT is especially applicable. The assumption that we are sampling from a normal distribution is a common assumption in statistics, which many times provides very good results. In all of the examples that we consider in this chapter, we will assume that the underlying distributions of interest are normal, or at least approximately normally distributed.

Proof of the CLT:

$$Z = \left(\frac{\overline{X} - \mu}{\sigma/\sqrt{n}}\right) = \left(\frac{\sqrt{n}}{\sigma}\right)\left(\frac{1}{n}\right)\left(\sum_{i=1}^{n}(X_i) - n\mu\right) = \left(\frac{1}{\sigma\sqrt{n}}\right)\left((X_1 - \mu) + \cdots + (X_n - \mu)\right)$$

The mgf of this Z is:

$$M_Z(t) = E(e^{tZ}) = \prod_{i=1}^{n} E\left(e^{\frac{t}{\sigma\sqrt{n}}(X_i - \mu)}\right) = \left[E\left(\sum_{k=0}^{\infty}\frac{t^k(X_i - \mu)^k}{\sigma^k(\sqrt{n})^k(k!)}\right)\right]^n$$

$$= \left[\left(\sum_{k=0}^{\infty}\frac{t^k E(X_i - \mu)^k}{(\sqrt{n})^k \sigma^k(k!)}\right)\right]^n$$

$$= \left[\left(1 + \frac{t^2}{2n}\right) + \left(\left(\frac{t^3 E(X_i - \mu)^3}{n^{3/2}\sigma^3(3!)}\right) + \left(\frac{t^4 E(X_i - \mu)^4}{n^2\sigma^4(4!)}\right) + \cdots\right)\right]^n$$

(The second part of this infinite sum goes to 0 as $n \rightarrow \infty$, since all the moments $E(X - \mu)^k$ are finite for $k \geq 3$)

Therefore, $M_Z(t) \rightarrow \left(1 + \frac{t^2}{2n}\right)^n = \left(e^{\frac{t^2}{2}}\right)$ as $n \rightarrow \infty$.

Therefore, $M_Z(t)$ converges to the mgf of the standard normal distribution. Recall that this is indeed the mgf of the standard normal distribution from chapter 8.

Therefore, since mgf's are unique when they exist, Z converges to the standard normal distribution.

In statistical applications, we generally don't know the variance σ^2, and hence σ. So we have to use S in place of σ. For large samples this is generally a good approximation, so we can extend the CLT to say that $Z = \sqrt{n}\left(\frac{\overline{X}-\mu}{S}\right)$ converges in distribution to the standard normal distribution for large samples.

(56) __HYPOTHESIS TESTING__

A large part of statistical inference for a single sample is the testing of competing hypotheses about the mean. The other main part is interval estimation (confidence intervals) which will be considered later. The first thing that we do after collecting our sample is to calculate $\overline{X}$ and S. Often we have a default value (μ_0) for the mean μ lacking any evidence to the contrary, and we would like to find evidence for the mean being less than (μ_0), or greater than (μ_0), or simply different from (μ_0). The hypothesis that μ is equal to the default value (μ_0) is called the null hypothesis H_0. The burden of proof is on finding strong enough evidence for the alternative hypothesis H_a. There are three alternative hypotheses H_a. So we test one of three situations:

H_0: $\mu = \mu_0$ vs. H_a: $\mu \neq \mu_0$
H_0: $\mu = \mu_0$ vs. H_a: $\mu < \mu_0$
H_0: $\mu = \mu_0$ vs. $H_{a:}$: $\mu > \mu_0$

The way that we test a pair of hypotheses is by computing a test statistic Z from $\overline{X}$,S,n, and μ_0, under the assumption that H_0 is true.

From the Central Limit Theorem, $Z = \sqrt{n}\left(\frac{\overline{X}-\mu_0}{S}\right)$ will be approximately normally distributed for large samples, regardless of the underlying distribution X. Z will be even more approximately normally distributed when we sample from normal distributions. It is common to assume that the underlying distributions that we are sampling from are normally distributed. So we will base our tests on Z, and later base our confidence intervals on Z. The computed value of Z for a particular sample will be in a certain set of numbers called the rejection region (RR) or in a certain set of numbers called the acceptance region (AR). Referring to the standard normal distribution curve in Figure 14, The AR and RR will be different for each of the three types of hypothesis testing situations. We have to first choose a so-called level of significance α for the test. Usually we choose α to be a small value like 0.10, 0.05, or 0.01. This α will determine the AR and the RR for a test of hypotheses.

For the first hypothesis testing situation, the AR will be the z-values corresponding to the middle $(1 - \alpha) \cdot 100\%$ of the Z distribution, and the RR will be the z-values corresponding to the lower $(\frac{\alpha}{2}) \cdot 100\%$ and the upper $(\frac{\alpha}{2}) \cdot 100\%$ of the Z distribution. We call these two different parts making up the RR the lower and upper tails of the Z distribution that together make up $(\alpha) \cdot 100\%$ of the Z distribution. These regions are disjoint and are determined by the level of significance α.

For the second and the third hypothesis testing situations, to determine the RR, we consider only the lower $(\alpha) \cdot 100\%$ of the Z distribution in the second case, or only the upper $(\alpha) \cdot 100\%$ of the Z distribution in the third case. So the RR is called lower-tailed or upper-tailed in the last two hypothesis testing situations.

We reject H_0 if $Z \in (RR)$. If H_0 is true, the $Pr(Z \in RR) = \alpha$ and the $Pr(Z \in (AR)) = 1 - \alpha$. So we will either reject H_0 and conclude that H_a is true, or we will fail to reject H_0 and accept that H_0 is true.

The main idea here is that if H_0 is true, the probability of getting an extreme Z value in the (RR) is so low that we should conclude that H_0 is not true and that H_a is more likely. If α is, for example, 0.05, then Z will be in the (RR) 5% of the time if H_0 is true, and we would be committing an error one out of every 20 times on average. If we choose $\alpha = 0.01$, then Z will be in the (RR) 1% of the time if H_0 is true, and we would be committing an error one out of every 100 times on average. The tradeoff for choosing a lower α is that we are less likely to find a false H_0 and a true H_a if in fact H_a is true. The error situation is summarized here:

There are two types of error that we can make.
(Type I error) We can reject H_0 when H_0 is true. The probability of this is our chosen α.
(Type II error) We can fail to reject H_0 when H_0 is false. The probability of this is called β, and it is generally more difficult to know because we don't know what the alternative value of μ is.

The value $(1 - \beta)$ is the probability of rejecting H_0 when H_0 is false, and we call this the power of the test. To compute the probability of a type II error β or the power of the test $(1 - \beta)$, we have to compute it with the assumption of a specific alternative value for μ. We will not compute β or $(1 - \beta)$ in any of the examples in this book. It will be assumed that when we have large sample sizes the probability of a type II error is low and the power of the test is high. All of the analysis in this book is for the case of large samples ($n \geq 30$). We should always try to work with as large a sample as possible when trying to make statistical inferences. There are methods for small sample situations. For example when using a small sample from a normal distribution X and σ is not known, and we estimate it with S, we can test the above hypotheses using a so-called T statistic, but we will not consider that here. Let's consider some examples.

Hypothesis Testing for a Mean

The following table will help us figure out the rejection region when we use the Z statistic to test our hypotheses for three common α-levels.

α- level	two-sided	lower tail	upper tail
0.10	$(-\infty, -1.645] \cup [1.645, \infty)$	$(-\infty, -1.285]$	$[1.285, \infty)$
0.05	$(-\infty, -1.96] \cup [1.96, \infty)$	$(-\infty, -1.645]$	$[1.645, \infty)$
0.01	$(-\infty, -2.575] \cup [2.575, \infty)$	$(-\infty, -2.33]$	$[2.33, \infty)$

Example 1: A chemist wants to find out if an experiment he is performing is creating less than 20 grams of sodium hydroxide on average with each time it is performed, at the $\alpha = 0.01$ level. We assume that the amount produced is normally distributed. He performs the experiment n = 40 times and gets an average $\overline{X}$ = 19.1 grams and standard deviation S = 0.8 grams. We wish to test:

H_0: $\mu = 20$ ($\mu_0 = 20$)   vs. H_a: $\mu < 20$

The chemist calculates:

$$Z = \sqrt{n}\left(\frac{\overline{X} - \mu_0}{S}\right) = \sqrt{40}\left(\frac{19.1 - 20}{0.8}\right) = -7.12$$

This Z is less than -2.33 and is highly significant. Therefore, the chemist rejects H_0 and concludes that less than 20 grams of sodium hydroxide is being produced on average each time that he performs the experiment.

Example 2: A scientist has been told that Arabian horses weigh 1230 lbs on average. He wants to test this claim to determine if they average either more or less than that at the $\alpha = 0.05$ level. We assume that the weights are normally distributed. He weighs n = 30 arabian horses and finds $\overline{X}$ = 1340 lbs, with S = 201 lbs. Therefore, to test:

H_0: μ = 1230 lbs   vs. H_a: $\mu \neq$ 1230 lbs

He calculates: Z = $\sqrt{30}$ ($\frac{1340-1230}{201}$) = 3.00

This Z is more than 1.96 and it is highly significant. Therefore the scientist concludes that the average weight of Arabian Horses is significantly different from 1230 lbs. It appears that on average they weigh more.

Example 3: A business manager at a large corporation computes that his 50 employees working under him are earning an average of $\overline{X}$ = $60,450 per year. He reads that the national average of employees in the same occupation is approximately normally distributed with mean salary of $59,900 per year, and standard deviation $3900. He wants to know if his employees are being paid significantly more than the national average. So he asks his company's statistical staff to test this. The chief statistician decides to test, at the α =0.05 level, the hypotheses

H_0: μ = $59,900   vs. H_a: μ > $59,900

So he computes Z = $\sqrt{50}$ ($\frac{60,450 - 59,900}{3900}$) = 1.00

Since the computed Z is not greater than 1.645, it is not significantly large. So the chief statistician reports to the manager that his employees are not making significantly more than the national average.

Hypothesis Testing for the Difference of two Means

Sometimes we have large samples from two independent normal

distributions X and Y, and we are interested in the difference of their means. In particular, we may be interested in whether or not the difference is zero, which would mean that the means are the same. From X, we have a sample of n_x observations with the statistics $\overline{X}$ and S_x. From Y, we have a sample of n_y observations with the statistics $\overline{Y}$ and S_y. The hypotheses that we would test would be one of these three types:

$$H_0 : \mu_x - \mu_y = 0 \quad \text{vs.} \quad H_a : \mu_x - \mu_y \neq 0$$
$$H_0 : \mu_x - \mu_y = 0 \quad \text{vs.} \quad H_a : \mu_x - \mu_y < 0$$
$$H_0 : \mu_x - \mu_y = 0 \quad \text{vs.} \quad H_a : \mu_x - \mu_y > 0$$

We will consider two examples and use $\alpha = 0.01$ as our level of significance.

Example 4: Two rival car companies claim that their hybrid models get the same gas mileage. 30 cars are tested from each of company X and company Y, and we assume that the gas mileage from each company's cars are normally distributed. So we have two independent distributions that we are sampling from. It is found that $\overline{X} = 41.2$, $\overline{Y} = 40.1$, $S_x = 2.9$ and $S_y = 3.0$. We wish to test the hypotheses:

$$H_0 : \mu_x - \mu_y = 0 \quad \text{vs.} \quad H_a : \mu_x - \mu_y \neq 0$$

The test statistic is Z = $\frac{(\overline{X}-\overline{Y}) - 0}{S}$,

where S^2 = Var($\overline{X} - \overline{Y}$) $= \frac{S_x^2}{n_x} + \frac{S_y^2}{n_y}$ and S $= \sqrt{\frac{S_x^2}{n_x} + \frac{S_y^2}{n_y}}$ is the standard deviation of ($\overline{X} - \overline{Y}$). Note that the variance of the linear combination of $\overline{X}$ and $\overline{Y}$, given by ($\overline{X} - \overline{Y}$), is just the sum of the variances since they are independent statistics. These results were outlined in chapter 7 on

Distributions and Expectation.

The calculated value of the test statistic Z is

$$Z = \frac{(41.2 - 40.1)}{\sqrt{(.2803) + (.3000)}} = 1.44 \ .$$

Since Z is within the AR = (-2.575, 2.575), we fail to reject H_0 and conclude that both company's hybrid models get the same gas mileage.

Example 5: In an opinion polling before an election, two independent samples are taken from the electorate. In the first sample, n = 600 potential voters are asked if they prefer candidate X. In the second sample, n = 600 potential voters are asked if they prefer candidate Y. The approval of candidate X was 53%, so $\overline{p_x}$ = 0.53. The approval of candidate Y was 47%, so $\overline{p_y}$ = 0.47. At the α = .01 level, we wish to test hypotheses to determine if candidate X has an advantage over Candidate Y. Note that these two statistics $\overline{p_x}$ and $\overline{p_y}$ are the binomial proportion statistics and for large samples are normally distributed from the CLT. So we will test:

$$H_0 : p_x - p_y = 0 \qquad \text{vs.} \qquad H_a : p_x - p_y > 0 \ .$$

The test statistic is Z = $\frac{(p_x - p_y) - 0}{S}$, where S = $\sqrt{\frac{\overline{p_x}(1 - \overline{p_x})}{n} + \frac{\overline{p_y}(1 - \overline{p_y})}{n}}$.

So Z = $\left(\frac{(.53 - .47)}{\sqrt{\frac{(.53)(.47) + (.47)(.53)}{600}}} \right)$ = $\frac{(.06)}{(.0288)}$ = 2.08

Since Z is not greater than 2.33, Z is not in the RR at the α = .01 level, so we fail to reject H_0. We are forced to conclude that candidate X does not have a significant lead over candidate Y. Note that at the α = .05 level we would conclude that candidate X has a significant lead.

(57) **CONFIDENCE INTERVALS**

The second main part of statistical inference is interval estimation for population parameters, which we call confidence intervals. Here we will consider confidence intervals for the mean μ and variance σ^2 of a distribution X. We will also include a Confidence Interval for the difference ($\mu_x - \mu_y$) between two distribution means. These things will involve the Z distribution and Chi-square distributions.

When n $\in [0, 100]$, z_n is the nth percentile of the Z distribution, meaning that the $Pr(Z \leq z_n) = n\%$. So for example, $z_{2.5}$ is a z-value such that the $Pr(Z \leq z_{2.5}) = 2.5\%$ and the $Pr(Z \geq z_{2.5}) = 97.5\%$ or, z_{90} is a z-value such that the $Pr(Z \leq z_{90}) = 90\%$ and the $Pr(Z \geq z_{90}) = 10\%$.

Similarly for the Chi-square distribution X with k degrees of freedom, $c_{(k),n}$ is the nth percentile of the Chi-square distribution with k df. So for example, $c_{(30),5}$ is a chi-square value from the Chi-square distribution X with 30 df, such that the $Pr(X \leq c_{(30),5}) = 5\%$ and the $Pr(X \geq c_{(30),5}) = 95\%$. Similarly, $c_{(34),80}$ is a chi-square value from the Chi-square distribution X with 34 df such that the $Pr(X \leq c_{(34),80}) = 80\%$ and the $Pr(X \geq c_{(34),80}) = 20\%$.

Some common values for the Z distribution:
$z_5 = -1.645$, $z_{2.5} = -1.96$, $z_{0.5} = -2.575$
$z_{95} = 1.645$, $z_{97.5} = 1.96$, $z_{99.5} = 2.575$

Confidence Interval for a Mean:

152

The Pr($z_\alpha < Z < z_{100-\alpha}$) = (100 - 2$\alpha$)%

So the Pr($z_{2.5} < Z < z_{97.5}$) = Pr($-z_{97.5} < Z < z_{97.5}$) = 95%.

For large samples from a distribution X with mean μ and variance σ^2, The Pr($-z_{97.5} < \sqrt{n}(\frac{\overline{X}-\mu}{S}) < z_{97.5}$) = 95%

So the Pr(($-z_{97.5}$)($\frac{S}{\sqrt{n}}$) < ($\overline{X} - \mu$) < ($z_{97.5}$)($\frac{S}{\sqrt{n}}$)) = 95%

So the Pr($\overline{X} - (z_{97.5})(\frac{S}{\sqrt{n}}) < \mu < \overline{X} + (z_{97.5})(\frac{S}{\sqrt{n}})$) = 95%

From this last statement, we call the interval,

[$\overline{X} - (z_{97.5})(\frac{S}{\sqrt{n}})$, $\overline{X} + (z_{97.5})(\frac{S}{\sqrt{n}})$] , a 95% confidence interval for μ.

By choosing different α-values, we can get a confidence interval of any desired confidence. Note that to construct this interval we have to conduct our experiment n times and calculate an $\overline{X}$ and an S. So this interval comes directly from the data and our conclusions from the CLT. It is not correct to say, with some specific calculated interval that the probability that μ is in that interval is 95%, instead we say that we have 95% confidence that μ is in that specific interval. The correct interpretation is that with repeated sampling and constructing confidence intervals from the above recipe, 95% of the time our intervals would contain the true but unknown μ. In the examples to follow we will for convenience always construct 95% confidence intervals.

Example 1: An agricultural scientist wants to get a 95% confidence interval estimate for the mean size of Ohio farms (in acres). He collects data on 30 randomly chosen farms and calculates $\overline{X}$ = 112 acres and S = 13.4 acres. He assumes that the farm sizes are normally distributed. So the calculated interval is,

$(112 - (1.96)(\frac{13.4}{\sqrt{30}}) , 112 + (1.96)(\frac{13.4}{\sqrt{30}})) =$
(107.2, 116.8).

Example 2: A study is conducted on the coin toss at the beginning of NFL football games. The coaches union wants to ensure that they are fair. They want a 95% confidence interval for the proportion of times that the home team won the toss. 100 football games from the last 5 years were randomly selected and it was noticed that 53 of the 100 coin tosses resulted in the home team winning the coin toss. Thus $\overline{X}$ = .53

and SD($\overline{X}$) = $\sqrt{\frac{(0.47)(0.53)}{100}}$ = 0.0499. So a 95% confidence interval for the true proportion p of times that the home team wins the toss (the union says that the confidence interval should contain 0.5 = 50% if the coin tosses have been fair) is:

[(0.53) - (1.96)(0.0499), (0.53) + (1.96)(0.0499)] = (0.432, 0.623).

Since this contains 0.5, the coaches union is satisfied with the result.

A Confidence Interval for the Difference Between Two means:

For large samples, the CLT says that the

$$Pr\left(-z_{97.5} < \frac{(\overline{X}-\overline{Y}) -(\mu_x-\mu_y)}{\sqrt{\frac{S_x^2}{nx}+\frac{S_y^2}{ny}}} < z_{97.5}\right) = 95\% .$$

Just as in the single sample case, we can solve this inequality for $(\mu_x - \mu_y)$ to get that the

$$\Pr\left((\overline{X}-\overline{Y})-(z_{97.5})\sqrt{\tfrac{S_x^2}{n_x}+\tfrac{S_y^2}{n_y}} < (\mu_x-\mu_y) < (\overline{X}-\overline{Y})+(z_{97.5})\sqrt{\tfrac{S_x^2}{n_x}+\tfrac{S_y^2}{n_y}}\right)$$

equals 95%.

So our large sample Confidence Interval for $(\mu_x-\mu_y)$ is

$$\left[(\overline{X}-\overline{Y}) \pm (z_{97.5})\sqrt{\tfrac{S_x^2}{n_x}+\tfrac{S_y^2}{n_y}}\right]$$

Example 3: Companies X and Y manufacture light bulbs and claim that the lifetimes of their products are the same. 50 bulbs from each company are tested and it is found that $\overline{X} = 1020$ hours and $\overline{Y} = 1110$ hours, with $S_x = 50.6$ and $S_y = 39.4$. Consumer regulators want a 95% confidence interval for $(\mu_y-\mu_x)$. The computed interval is:

$$((1110\text{-}1020) \pm (1.96)\sqrt{51.2072+31.0472})$$

which is $(90 \pm (1.96)(9.069))$, or $(72.23, 107.78)$.

This is a 95% Confidence Interval for the hypothetical but unknown difference $(\mu_y-\mu_x)$. Examining this interval, it looks as if company Y light bulbs have a much longer lifetime than company X light bulbs.

A confidence Interval for the Variance:

If we conduct an experiment a large number of times n, and the distribution X being sampled from is normally distributed. We calculate $\overline{X}$ and S^2, and we know that $\left(\tfrac{(n-1)S^2}{\sigma^2}\right)$ is Chi-square distributed with approximately (n) df.

So the $\Pr(c_{(n),2.5} < (\frac{(n-1)S^2}{\sigma^2}) < c_{(n),97.5}) = 95\%$

So the $\Pr((\frac{(n-1)S^2}{c_{(n),97.5}}) < \sigma^2 < (\frac{(n-1)S^2}{c_{(n),2.5}})) = 95\%$

This leads us to a 95% confidence interval for σ^2,

$[(\frac{(n-1)S^2}{c_{(n),97.5}}) , (\frac{(n-1)S^2}{c_{(n),2.5}})]$.

Example 4: Suppose an airline needs a 95% confidence interval for the variance σ^2 and the standard deviation σ of the flight time for a certain routine flight from Washington to Tampa. It is assumed that the flight times are normally distributed. The company collects 100 flight times, randomly chosen from the last 250 flights, and finds $\overline{X}$ = 95.3 minutes and S = 8.4 minutes. We will use the values $c_{(100),2.5}$ = 74.22, $c_{(100),97.5}$ = 129.56, which can be obtained from published tables of the Chi-square distribution. So the 95% confidence interval for the true but unknown variance σ^2 is:

$(\frac{(99)(70.56)}{(129.56)} , \frac{(99)(70.56)}{(74.22)}) =$

(53.92, 94.12).

By taking square roots we can get a 95% confidence interval for the true but unknown standard deviation σ: (7.34, 9.70).

Printed in the United States
By Bookmasters